AGRIBUSINESS REFORMS IN CHINA

The Case of Wool

Dedicated
to
Jill and Robyn

AGRIBUSINESS
REFORMS
IN CHINA

The Case of Wool

John W. Longworth and Colin G. Brown

Department of Agriculture
The University of Queensland
Australia

CAB INTERNATIONAL

In association with:

ACIAR
The Australian Centre for International Agricultural Research

HD
9906
.C62
L65
1995

CAB INTERNATIONAL
Wallingford
Oxon OX10 8DE
UK

Tel: +44(0)1491 832111
Telex: 847964 (COMAGG G)
E-mail: cabi@cabi.org
Fax: +44(0)1491 833508

A catalogue record for this book is available from the British Library.

ISBN 0 85198 951 9

Published in association with:

The Australian Centre for International Agricultural Research (ACIAR)
GPO Box 1571
Canberra
ACT 2601
Australia

Tel: (06) 248 8588
Fax: (06) 257 3051
Telex: AA 62491
Telecom Gold/Dialcom: 6007: IAR001

Printed and bound in the UK at the University Press, Cambridge.

Contents

Appendices

List of Tables

List of Figures

Foreword

As the huge agribusiness sector in China modernises and becomes more integrated into international markets, there will be massive adjustment problems to be overcome within China. At the same time, participants in world markets will need to take more account of what is happening in China. China has already emerged as a major participant on the international wool scene. It has also drastically restructured its domestic wool market. This volume suggests ways in which Chinese policy-makers should continue to modernise their wool sector, and provides a detailed case study for foreigners who want to understand China's agribusiness sector.

The research upon which this book is based was funded primarily by The Australian Centre for International Agricultural Research (ACIAR) which promotes collaborative research on problems of major significance in the rural sector of less developed countries. The aim is to encourage Australian researchers to assist colleagues in these countries undertake research which might not be feasible otherwise owing to a lack of either expertise or research resources. In selecting projects, ACIAR concentrates both on fields of research in which Australian scientists have a comparative advantage and on issues of potential relevance to Australian rural industries. This book is one outcome of a project which satisfies both these criteria: Australian agricultural economists have a long history of involvement with wool economics research and the Chinese wool industry is of great commercial interest to the Australian wool industry.

An Australian team of agricultural economists, based at The University of Queensland, collaborated with researchers from two Chinese institutions—the Institute of Rural Development (Chinese Academy of Social Sciences) and the Institute of Agricultural Economics (Chinese Academy of Agricultural Sciences)—on a project entitled "Economic Aspects of Raw Wool Production and Marketing in China". The study commenced in mid-1989 and concluded with a final workshop in early 1994. Publications arising from the research include two major books and a number of articles in Chinese together with three books (including this volume), a technical report and many professional papers in English. While ACIAR resources played a major catalytic role in linking the various groups of researchers, all parties contributed both human and financial resources. The project benefited greatly from the continued support and direct involvement of the Directors of both Chinese Institutes.

In addition to the ACIAR project, there was a related but separate study conducted by Colin Brown as part of an Australia–China Agricultural Cooperation Agreement mission to China in 1992 to investigate wool processing. The principal findings of this study have also been incorporated in this volume.

The major economic reforms which have transformed the rural sector of China since the late 1970s have been subjected to close scrutiny both by Chinese and foreign scholars. Seldom, however, have western economists with the necessary technical skills and experience to appreciate the subtleties of a particular marketing system, had an opportunity to study one part of China's huge agribusiness sector in such detail over such a long time period. John Longworth is a qualified wool classer and sheep judge as well as an agricultural economist. This book, represents a comprehensive analysis of the impact of the various reforms affecting the wool market both in an economic sense and in relation to technical detail. It examines the consequences for the various interest groups and marketing institutions not only of general reforms such as the move to more open markets, but also of commodity specific reforms such as the introduction of new grading standards.

Kenneth M. Menz
Research Program Coordinator (Economics and Farming Systems)
Australian Centre for International Agricultural Research, Canberra

Acknowledgements

Unravelling the intricacies of the long and complex wool marketing and processing chain in China is a complex task. It would have been impossible for this book to have been written without the cooperation and enthusiastic support of a plethora of individuals and organisations in China. These included individual sheep herders, State farm managers, local SMC buyers, county scouring-plant managers, officials at large State-owned textile mills, and many government officials at all levels. While the contributions of all these people must remain anonymous, the authors gratefully acknowledge their invaluable assistance. It is hoped that the conclusions and suggestions contained in this volume contribute to advancing reforms in the Chinese wool sector and hence to improving the lot of the people whose livelihoods depend upon wool.

Much of the research on which this book is based was made possible by funding provided by the Australian Centre for International Agricultural Research (ACIAR). The collaborative research project supported by ACIAR brought together researchers from The University of Queensland and two Chinese institutes—the Institute of Agricultural Economics (within the Chinese Academy of Agricultural Science) and the Institute for Rural Development (within the Chinese Academy of Social Science). The Chinese members of the research team not only contributed a great depth of local knowledge and contacts but also served as interpreters and attended to the many mundane logistical aspects of the research in China. In particular, the authors would like to acknowledge the commitment to the project of the Directors of the two collaborating Chinese institutions, Professor Niu Ruofeng and Professor Chen Jiyuan. Other senior Chinese scientists working on the project included Zhang Cungen, Lin Xiangjin, Zhou Li, Liu Yuman, Xu Ying and Du Yintang. All of these people have made major contributions to our understanding of the Chinese wool industry.

Fieldwork for the ACIAR project commenced in China in August 1989, shortly after the political crisis in May/June of that year. From 1989 to 1993, Australian and Chinese scientists worked together for up to three months each year travelling widely in northern and north western China to gain first-hand impressions and to collect primary data. During this fieldwork the researchers were greatly assisted by the cooperation of many government officials at various levels in the Ministry of Agriculture, Ministry of Textile Industry and Ministry of Commerce. In particular, the authors would like to thank the many individuals working for the central, provincial, prefectural and county animal husbandry bureaus within the Ministry of Agriculture. Much of the fieldwork was only feasible because these people made themselves and their resources available.

In 1992, Colin Brown participated in a mission to investigate wool processing in China financed by the Australian and Chinese Governments under the Australia–China Agricultural Cooperation Agreement (ACACA). The members of this mission built on earlier research and insights gained from the ACIAR project. Special attention was devoted to studying the problems facing the domestic-market oriented mills, particularly those located up-country in the three major wool-growing provinces of IMAR, Gansu and XUAR. During the ACACA mission, textile industry, mill fibre inspection and SMC officials in these three provinces were extremely helpful and provided invaluable information. Zhao Bing of the China International Exchange Association of Agriculture within the Ministry of Agriculture is thanked for his guidance and interpreting during the mission. Special mention must also be made of He Zheming and Li Jun from the textile industry corporations of IMAR and XUAR respectively, who enthusiastically assisted both authors on a number of occasions.

Apart from the various wool industry participants in China, other outside organisations and individuals have also contributed to this work. The contributions of the International Wool Secretariat, the Australian Wool Corporation and the Australian Wool Research and Promotion Organisation through John Sheung, Carolyn Stewart, Wang Cheng Guang, Chris Wilcox, Peter Hawley and others is gratefully acknowledged. The authors would also like to thank the large number of individuals from the Chinese, Australian and New Zealand wool and wool textile sectors who participated in the ACIAR sponsored workshop on the Chinese wool industry held in Wellington, New Zealand in February 1994, for the free exchange of information and the cross fertilisation of ideas which occurred at that meeting.

Two other members of the Australian research team, Greg Williamson and Ross Drynan, played major parts in the fieldwork in China on which this book is based. We wish to thank Greg and Ross for their generosity in allowing us to incorporate ideas and data recorded in their fieldnotes without specific acknowledgment. Greg Williamson, in particular, has made a major contribution since he was the only Australian scientist employed fulltime on the ACIAR project.

The authors wish to acknowledge the support of their employer, The University of Queensland. The University, through its Department of Agriculture, made available office space and other research support services to complement the direct financial support provided by ACIAR. It must also be noted that ACIAR has also contributed to this book through the editorial and other efforts of Peter Lynch and Ken Menz. We thank these people for their encouragement and continued support.

In 1991, Alistair Watson and Zheng Dahao served as external members on the panel which reviewed the ACIAR project on which this book is primarily based. The authors wish to thank these two people not only for the favourable outcome of the review which allowed the project to be extended, but also for their many thoughtful suggestions. We have also had the benefit of feedback from Al Watson on an earlier draft of this volume.

In the preparation of this book and throughout the ACIAR project, the authors have been extremely fortunate to have had available the typing, typesetting and general secretarial skills of Debbie Noon. Deb is the most patient, meticulous and dedicated preparer of manuscripts one could ever find. Thank you Deb for maintaining your interest and enthusiasm for so long. You have played an invaluable role in the production of this book (and all the other English language material emanating from the ACIAR project). Colleen Moss is another person who has contributed significantly since she has edited the whole manuscript at least twice. Colleen's grasp of the finer points of English grammar and punctuation has improved the readability of the final product immeasurably.

Apart from the time spent in China on fieldwork in past years, the preparation of this book has severely encroached on family time during the last twelve months. We wish, therefore, to thank both Jill and Robyn for their patience and their support.

Finally, with the help of all those mentioned above and many others too numerous to name, we have endeavoured "to get it right". The errors which remain, however, are entirely our responsibility.

John W. Longworth   December 1994
Colin G. Brown

About the Authors

John Longworth is Professor of Agricultural Economics and Pro-Vice-Chancellor (Social Sciences) at The University of Queensland. Prior to commencing an academic career, he became a qualified wool-classer and was employed by GRAZCOS (at that time, the world's largest shearing contractor). This experience, together with John's life-long involvement in commercial sheep-and-wool production on his family farm enables him to appreciate fully the practical implications of wool marketing reforms in China.

At the same time, John Longworth is a former President of the International Association of Agricultural Economists and a Fellow of the Academy of the Social Sciences in Australia. His research on the socio-economic aspects of Japanese rural policy, reported in the book *Beef in Japan* (1983), led to similar investigations in China. Since 1986, John has visited China many times. He has travelled widely in the pastoral wool-growing areas of China conducting fieldwork. In addition to this volume, John has published a number of research papers and three other books on China: *China's Rural Development Miracle* (1989); *The Wool Industry in China* (1990); and *China's Pastoral Region* (1993) (with Greg Williamson).

Colin Brown is Lecturer of Agricultural Economics in the Department of Agriculture at The University of Queensland. Colin joined the ACIAR Project on which this book is primarily based in 1990 and much of his fieldwork and research has focused on marketing and processing aspects of wool in China. These interests were developed further in 1992 as part of an Australia–China Agricultural Cooperation Agreement mission on wool processing in China. Since then, Colin has participated in a number of workshops and seminars in this field.

Apart from his involvement with wool marketing and agribusiness reforms in China, Colin has conducted extensive research on agribusiness reforms in other countries. As a research officer with the Danish Institute of Agricultural Economics in the latter half of the 1980s and again in 1993, Colin conducted extensive research and published widely on reforms to the Common Agricultural Policy of the European Community. In the 1980s, he also undertook research into the spatial organisation, labour arrangements and other impediments to efficiency in the Australian cattle slaughtering industry. Although the research in Europe and in Australia relates to entirely different agribusiness environments, the experience gained in examining the close linkages existing between policy and marketing issues enable Colin to comment authoritatively both on the specific wool marketing reforms as well as the more general agribusiness reforms in China.

Kenneth M. Menz
Research Program Coordinator (Economics and Farming Systems)
Australian Centre for International Agricultural Research, Canberra

Glossary

Abbreviations

ABC	Agricultural Bank of China
ACIAR	Australian Centre for International Agricultural Research
AHB	Animal Husbandry Bureau
AHICC	Animal Husbandry Industrial and Commercial Company
AIDAB	Australian International Development Assistance Bureau
AWRAP	Australian Wool Research and Promotion Organisation
BWTRI	Beijing Wool Textile Research Institute
CCP	Chinese Communist Party
CHINATEX	China Textile Import and Export Corporation
FIB	Fibre Inspection Bureau
FYP	Five-Year Plan
GATT	General Agreement on Tariffs and Trade
IMAR	Inner Mongolia Autonomous Region
IWS	International Wool Secretariat
MOA	Ministry of Agriculture
MOC/MOIT	Ministry of Commerce/Ministry of Internal Trade
MOFERT/MOFTEC	Ministry of Foreign Economic Relations and Trade/ Ministry of Foreign Trade and Economic Corporation
MOTI/CNTC	Ministry of Textile Industry/ China National Textile Council
PCC	Production and Construction Corps
PRC	People's Republic of China
SMC	Supply and Marketing Cooperative
TIC	Textile Industry Corporation
TRI	Textile Research Institute
WTO	World Trade Organisation
XUAR	Xinjiang Uygur Autonomous Region

Units

yuan	unit of money (¥) (see below)
μm	micron ($1\mu m = 10^{-6}m$)
spinning count	traditional wool trade term to describe the fineness of wool with lower values indicating coarser wool and higher values signifying finer wool

Currency

The Chinese currency is called the Renminbi (RMB) but the unit of money is the yuan. The exchange rate yuan per US dollar since 1980 has been as follows:

1980	1.50	1985	2.94	1990	4.78
1981	1.70	1986	3.45	1991	5.32
1982	1.89	1987	3.72	1992	5.52
1983	1.98	1988	3.72	1993	5.80
1984	2.32	1989	3.76	1994	8.60

Source: Bank of China.

Chapter 1

Soft Gold

Even in the heady world of Chinese commodity and product markets, the wool market stands out. There are few sub-systems in the Chinese rural sector which provide so many rich illustrations of how agribusiness is blossoming in China and how recent reforms are influencing this growth. From rudimentary markets based on a nomadic production system in the 1950s, the Chinese wool market has evolved to such an extent that in the 1990s it represents an important component of the world market for wool.

The strong growth in both domestic and international demand for wool-based products manufactured in China has seen that country become dominant in international markets, both as an importer of raw and semi-processed wool and as an exporter of wool textiles, garments and yarns. With the deregulation of the domestic wool (and wool product) markets in the early 1990s, the Chinese wool market now not only links domestic wool growers with domestic processors as in the past, but also integrates the whole Chinese wool-growing and processing industry into the international wool trade much more than previously. However, the exposure of Chinese wool growers to international market forces has the potential to create major difficulties for the Government of the People's Republic of China (PRC).

Wool growing is an economic activity of great political and strategic significance in the PRC. After Australia and New Zealand, China has the third largest sheep flock in the world. The majority of these sheep, and especially the animals which produce apparel wool, are raised in remote, economically backward, pastoral areas by people who belong to minority nationalities. Many of these sheep herders are among the poorest people in the country and wool represents their only "cash crop". Indeed, wool is so valuable to these people that they often refer to it as "soft gold". If these minority groups are to participate in national economic development, it is imperative that their income from wool is not only maintained but improved. Furthermore, the political and strategic sensitivity of wool growing is heightened by the traditional ethnic and cultural links between the sheep-raising minorities and the peoples who inhabit the countries to the north and west of China.

Wool, therefore, has moved to centre stage in Chinese policy-making. It has proven to be one of the major obstacles to China rejoining the General Agreement on Tariffs and Trade (GATT) and becoming a founding member of the proposed new World Trade Organisation (WTO). The Chinese government has sought to retain the right to limit wool imports by quotas if necessary to protect the incomes of the minorities who grow wool. Of course, the Cairns Group of countries (led by Australia) have vigorously opposed the granting of such a concession to China.

Wool growing and marketing in China are not well understood either by foreigners or by many Chinese policy-makers. The production aspects of the Chinese wool industry were recently documented by Longworth and Williamson (1993). The aim of this book is to complement the Longworth and Williamson volume by providing a comprehensive analysis of the Chinese wool-marketing system. An understanding of this system is critical for wool traders and others interested in likely future developments in raw-wool production in China, in the Chinese wool textile industry, and in China's role in international markets for wool and wool products.

From an agribusiness perspective, however, a detailed appreciation of the Chinese wool market is not just important for wool and wool product traders. The recent reforms have created business opportunities along the length of the marketing chain as traditional institutions endeavour to respond to the new commercial environment and as new agencies enter the market. Whether it be the need for environmentally sound waste-disposal technology for wool scours or improved fibre inspection methods, the Chinese wool market offers many new market niches for foreign agribusiness entrepreneurs. This book highlights some of these commercial opportunities but does not attempt an exhaustive listing. Rather the objective has been to highlight how and why the recent wool market reforms have created the need for new agribusiness technologies and services. In this sense, the study is of considerable generality since it provides a good example of how the huge agribusiness sector in China is being modernised.

1.1 China Through the "Wool Window"

Clearly, wool marketing in China is of great interest to a wide spectrum of people involved with wool trading and with wool-connected agribusiness activities. But an examination of the reforms which have restructured the distribution and exchange of wool in China also offers a unique "window" through which to observe the profound changes which have transformed Chinese society since the initial official sanctioning of the economic reform movement in late 1978. An investigation of wool-marketing reforms illustrates many of the major problems and issues confronting contemporary Chinese society and its interaction with the world community. From social and cultural to environmental and development issues, from regional security to international relations, from domestic economic policy to world trade negotiations, Chinese wool marketing impacts on them all. Consider the following:

• Wool is produced in the vast and remote pastoral region of China. Despite covering some two-thirds of the Chinese land mass, this region is largely unknown outside China as most attention focuses on the east coast provinces. The region is of great strategic importance to China as it includes many sensitive border areas. New trade opportunities (or perhaps the reopening of some ancient trade routes) have emerged recently. For instance, the major wool-producing province of Xinjiang in the north west corner of China has become the economic epicentre of a vast area of Central Asia. Sharing its borders with Mongolia, Afghanistan, Pakistan, Kashmir and four of the previous Soviet Republics, it has become a hub of economic and entrepreneurial activity as the neighbouring countries seek to work through changing political systems and depressed economic conditions.

• Wool traditionally was produced in China by nomadic herders of the many tribes and nationalities that inhabited this part of Asia. Although production systems have changed and there has been substantial immigration of Han Chinese into these pastoral areas, wool production is still largely conducted by minority nationalities in regions where minorities account for a substantial proportion of the population. Consequently, many of the administrative regions are designated as autonomous units and the local administrations have a certain degree of autonomy relative to the Central government. A study of how wool marketing has evolved provides some good insights into the policies directed towards these autonomous regions and the impact of the greater degree of local autonomy on economic and social development.

- Not only are the wool-growing areas populated by minority nationalities but also these people are among the poorest in China. The widespread poverty within the pastoral region has not only resulted in specific policies such as the various poverty alleviation measures but it has also inhibited the modernisation of marketing systems for products such as wool as well as the ability of the region to grasp new economic opportunities.

- The resource on which wool production is based, namely the rangelands, is rapidly degrading. Like other arid and semi-arid regions in poor areas around the world, China's pastoral region is in the grip of a vicious cycle of increasing human populations leading to pressure to increase livestock stocking rates to maintain incomes. This, in turn, leads to greater rangeland degradation, reducing the capability of the rangelands to support the increasing livestock populations and the human populations which rely on them. There is an urgent need to move to a more sustainable development strategy for the wool-growing areas of China.

- Wool-marketing reform in China graphically highlights some of the transitional problems with general market and economic reforms. The "wool wars", which led to a chaotic market environment, were but one example of the so-called commodity wars that broke out in the second half of the 1980s. As a result of the problems created by the "wool wars", wool-marketing reform has proceeded in a cautious stop-start fashion as policy-makers respond to adverse unforeseen effects in a characteristically Chinese step-by-step approach to policy change.

- Wool has a lengthy marketing channel from production through various stages of wool processing (scouring, topmaking, yarn and fabric making) to final textile and garment manufacture. The Chinese wool industry is involved in all these stages with both domestic and international linkages. Consequently, it provides useful insights into how reforms at one stage can impact on the other stages. For instance, the impact of the introduction of the household production responsibility system in the production sector has had a major influence on wool marketing and vice versa.

- Many wool-marketing reforms undertaken illustrate both the opportunities and potential problems involved in transforming the Chinese agribusiness sector from being comprised of centrally planned and administered marketing systems to a sector of relatively free markets. One interesting example is the development of Chinese wool auctions since 1987. Another is the development of direct selling between wool producers and wool processors. The wool market illustrates clearly the difficulties encountered in weaning industries off highly regulated marketing arrangements.

- The Chinese wool-processing industry contains a cross-section of Chinese manufacturing industry. Numerous large, State-owned or State-controlled mills exist and these factories are typical examples of the State enterprises in China which are currently under great pressure to reform. However, wool textile mills were also one of the most popular forms of rural-township enterprise that sprang up during the 1980s. These township-enterprise mills suffered major difficulties during the 1989 to 1992 wool depression. In addition, many new early-stage processing plants were established in wool-growing areas in the late 1980s, creating excess capacity and quality control problems.

- Because of the importance of wool imports and wool textile exports, the Chinese wool market also illustrates many elements of Chinese foreign trading arrangements. For example, the internal battles fought between the central trade agencies such as the Ministry of Foreign Economic Relations and Trade (MOFERT) and regional trade organisations have been especially notable in the wool market.

- Joint ventures have also been a feature of wool-processing mills and other wool industry organisations. In particular, the joint ventures have been used in the recent past to circumvent restrictive wool import arrangements. More generally, joint ventures have been a major means by which equipment and technology have been upgraded and they have helped the Chinese industry keep abreast of fashion and market developments overseas.

1.2 Scope of the Book

The comments in the previous section suggest a potentially very broad canvas but a study of the kind reported in this volume must, of necessity, have a clear focus. This book, therefore, concentrates on the raw-wool end rather than the wool-product end of the marketing chain. Previous studies have paid more attention to both the domestic consumer market and the impact of Chinese exports on world markets for wool textiles, garments and yarns (e.g. Anderson and Park, 1988; Garnaut *et al.*, 1993; IWS, 1994). Neither of these topics is considered directly in this book. Instead, the emphasis is on how wool, both domestically grown and imported, reaches the mills.

No specialised knowledge of wool nor any detailed prior appreciation of conditions in China are assumed. However, throughout the text references are made to relevant literature which will provide the specialist with further detail.

One of the features of this study is that it is based on extensive fieldwork in wool-growing areas of China and as a result many specific locations are mentioned. The broadest such geographic unit to which reference is made is the pastoral region defined as the 12 provinces indicated in Fig. 1.1. Three of these provincial-level administrative units—the Xinjiang Uygur Autonomous Region (XUAR), Gansu, and the Inner Mongolia Autonomous Region (IMAR)—receive special attention. These three provinces are discussed in considerable detail in Longworth and Williamson (1993, pp.75–164). The next administrative unit below the province is the prefecture and several important wool-growing prefectures are mentioned. The next step down in the governmental hierarchy is the county and many important wool-producing counties are also named in this book. Fig. 1.2 indicates the general location of the prefectures and counties to which frequent reference is made in the following chapters. The 10

Fig. 1.1 The 12 Pastoral Provinces of China

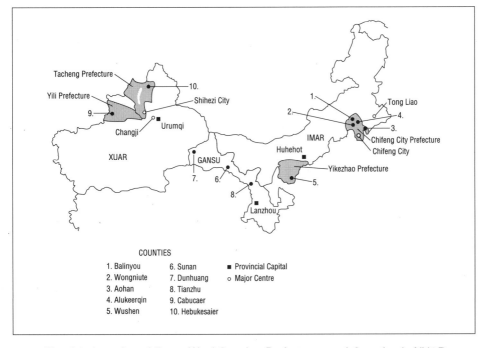

Fig. 1.2 Location of Some Wool-Growing Prefectures and Counties in XUAR, Gansu and IMAR

counties, the general location of which are indicated in Fig. 1.2, are presented as case studies in Longworth and Williamson (1993, pp.167–295). The lowest level of formal government in China is the township, which corresponds with the old commune level of administration. Township governments are responsible for administrative villages which may embrace one or more natural villages. No specific townships or villages are given any emphasis in this book, although the research on which this volume is based included a large number of interviews at the township, village and even household level.

1.3 Sources of Information

Wool marketing in China involves a number of distinct vertical hierarchies within the Chinese bureaucracy. These hierarchies frequently have parallel administrative structures at each level of government from the national level down to the county and even to the township level in some cases (Longworth, 1990, pp.3–5). The official horizontal linkages between these vertically organised bureaucracies are usually weak and sometimes non-existent. There is no alternative but to access each bureaucracy as close to the top as possible and then work down the structure collecting and cross-checking information at each level. Only in this way is it possible to gain an appreciation of the real situation within the ambit of responsibility of each institution or organisation. Once first-hand data have been collected and verified in this way, it is possible to construct a system-wide view by integrating and synthesising across the rigidly defined areas of influence assigned to each vertically organised bureaucracy. Since Chinese administrators and researchers rarely adopt such a comprehensive approach, they frequently fail to identify important inconsistencies or policy failures even though, once pointed out, these problems may appear to be self-evident. In this regard, this book not only draws together data and information from a great many diverse sources but it also presents a system-wide view which was not previously available, even in Chinese. Indeed, relevant Chinese language literature is extremely sparse and is usually too general or anecdotal to be of any real use. Official statistics available in Beijing are also of limited value owing to the high degree of aggregation and because the data are often open to numerous possible interpretations.

Consequently, apart from making the most of all the available secondary sources, this book is based principally on primary information obtained from interviews and surveys conducted as part of a Sino–Australian collaborative research project aimed at identifying the constraints to further development of the Chinese wool-growing industry. This project was organised under the auspices of the Australian Centre for International Agricultural Research (ACIAR). Fieldwork was conducted in each year from 1989 to 1993 and so spanned a period characterised by momentous changes in Chinese wool marketing and Chinese society in general.

Interviews were conducted at relevant administrative levels in all the vertical hierarchies concerned with the marketing of "soft gold". Particular attention was focused on the Animal Husbandry Bureaus (AHBs), Supply and Marketing Cooperatives (SMCs), Fibre Inspection Bureaus (FIBs), Textile Industry Corporations (TICs) and Price Bureaus. However, officials and managers of many other organisations such as textile mills, wool research institutes, university departments, and local government policy divisions, were also interviewed. Details of the ACIAR project are available in Longworth and Williamson (1993, pp.9–24).

Apart from the ACIAR collaborative research project, fieldwork was also conducted in 1992 as part of an Australia–China Agricultural Cooperation Agreement mission on wool processing in China. This investigation proved invaluable by providing supplementary information on issues related to wool processing and wool importing. Details of this mission and some of the primary information obtained are reported in Brown (1993).

1.4 Organisation of the Book

Taken together, the topics covered in this book provide not only a comprehensive picture of the evolution of the market for raw wool in China but also examples of how the whole agribusiness sector in China is being modernised. The text is organised into more-or-less self-contained chapters and appendices that may be read separately depending upon the special interests of the reader. In this regard, while the appendices need not concern the general reader, they are a feature of the book that will appeal to the wool specialist since they provide translations of previously unavailable wool standards and wool quality-control regulations.

While each chapter has its own story to tell, Chapters 2 and 3 contain important background information that provides a context for the discussions in the remainder of the book. In particular, Chapter 2 presents an overview of the wool industry in China. First, the agribusiness environment or marketing channels and associated institutions are described and the various important features of the market system to be taken up in subsequent chapters are identified. Of course, the present marketing arrangements reflect the special characteristics of both wool growing and wool processing in China. Therefore, these two sectors of the overall wool industry are briefly described with emphasis on identifying those special aspects of growing and processing which have implications for the modernisation of the marketing system.

While Chapter 2 presents an essentially cross-sectional picture of the present situation, Chapter 3 provides a longitudinal or historical perspective. It briefly describes the major political and economic events which have influenced the development of the Chinese wool industry since the founding of New China in 1949. As with the rest of the Chinese economy, the wool industry has been through some extremely turbulent times and these experiences have influenced greatly the development of the industry in recent years.

The most fundamental reform in relation to the marketing of wool has been the switch from administered pricing to a nominally free market situation. However, before this deregulation of the price-setting process is discussed in Chapter 5, it is necessary to understand how wool is graded in China. Therefore, Chapter 4 explores the surprisingly diverse and detailed industry standards which define the grading systems currently in use. Some of the major difficulties which still exist in relation to grading and pricing, and which are discussed in Chapter 5, relate to the inability of the marketing system to provide appropriate fibre inspection and testing services. These issues are addressed in Chapter 6. While, as already stated, each chapter in this book can be read separately, Chapters 4 to 6 are closely interconnected and really need to be considered together.

The SMCs are a massive agribusiness network supplying a wide range of farm inputs and marketing many different farm products, of which wool is perhaps one of the most important. While still retaining great political and economic power in relation to wool marketing, the SMCs have been under increasing pressure from the recent reforms to modernise their operations. The nature of the emerging competitive environment faced by the SMCs and their reaction to the new situation are among the issues considered in Chapter 7.

Wool auctions could be seen as short-lived novelties in the overall context of wool-marketing reform. Nevertheless, as the analysis in Chapter 8 demonstrates, the experimentation with auctions illustrates the diverse pressures and constraints impinging on the reform process. In addition, the data discussed in relation to the auctions provide a rare picture of the physical characteristics of some of the best wool produced in China. These details have not been available previously, even in China.

Another major development in the latter part of the 1980s has been the establishment of a large number of new early-stage processing plants in remote wool-growing counties. Chapter 9 investigates the implications of this development and highlights the major potential difficulties which these plants have created.

The next two chapters are concerned with the textile industry. In Chapter 10 the general problems currently facing the Chinese wool textile manufacturing sector are evaluated, with particular emphasis on the difficulties the domestic raw-wool-marketing system creates for manufacturers. Chapter 11 concentrates on the flow of imported wool to the mills. There have been major changes in relation to the importation of raw materials for the Chinese wool textile industry in the 1990s. Some of these changes have been the result of economy-wide reforms but the most important have been almost unofficial adjustments to how wool importation is administered. There is an emphasis in Chapter 11 on the need to consider the special needs of the mills producing for the domestic market (especially the up-country mills) if the full potential of the Chinese wool market is to be realised.

The final chapter highlights some of the major lessons for the future which emerge from the detailed discussion in the earlier chapters. Particular emphasis is placed on the enormous amount of new economic activity which the emerging agribusiness sector in China is generating. While in value terms wool is a relatively minor product in the totality of the Chinese rural sector, it is a commodity which has rapidly increased in economic, strategic, political and trade-policy significance in recent years. The growing importance of wool in China stems both from domestic considerations and from the sudden emergence of the Chinese wool textile industry as a dominant player in the international markets both for raw and semi-processed wool and for wool textiles. This detailed case study of Chinese wool-marketing reforms, therefore, is a timely contribution not only towards a comprehensive understanding of the Chinese wool scene, but also towards a better appreciation of how economic reforms in China are creating a massive new agribusiness sector.

Chapter 2

Sheep's Side to Consumer's Back

Converting a fleece obtained from a sheep in the remote pastoral region of China into a suit worn by a Shanghai business executive is a long and complicated process. This reflects not only the nature of wool and wool processing but also the unique features surrounding the agribusiness environment for wool in China. Specific elements of this environment are examined in subsequent chapters, with an emphasis on the raw-wool end rather than wool-product end of the marketing system. The purposes of this chapter are threefold: first, to provide a perspective on how the topics discussed in later chapters relate to the overall wool-marketing system in China and to each other; second, to highlight salient aspects of wool growing in China as background to the later discussions; and third, to present an outline of the Chinese wool-processing sector. In these latter two sections, the aim is to demonstrate how conditions in the growing and processing sectors impinge on marketing.

2.1 Agribusiness Environment for Wool and Key Institutions

The agribusiness reforms which are transforming wool marketing in China need to be seen in totality. That is, the impact of any one set of changes needs to be assessed in the context of the total marketing system rather than in isolation. This is a major theme of this book. Therefore, this section begins with a brief outline of traditional pre-reform marketing arrangements and then, with the aid of a simple diagram, the various institutions concerned with the marketing of wool are introduced briefly. The objective is to provide an overview of the total system that links growers to processors (and ultimately to consumers). At the same time, it is demonstrated how the topics discussed in the various chapters of this book relate to one another and to the total marketing system.

2.1.1 Traditional Wool-Marketing Channel

Traditionally, there was only one major marketing channel for wool in China. Under the centrally planned production and distribution system operated by the People's Republic of China (PRC), the Supply and Marketing Cooperatives (SMCs) were designated as the State procurement agency. That is, all wool grown to satisfy production quotas was purchased at State-determined prices by the SMCs. The SMCs delivered the wool to mills designated in the plan to process it. The mills paid the SMCs the procurement price plus a fixed mark-up. The mills delivered their products to State-controlled distribution networks at prices which were calculated on the basis of the mill-door price the mills paid for their raw materials plus an approved mark-up. Finally, State shops sold the wool-based consumer products to customers at fixed prices.

In addition to the State-controlled marketing channel just described, over-quota production was sometimes sold through local free markets, but the volume of wool handled in this way was always small, as these markets only serviced the needs of local communities. In the major wool-growing areas, the SMCs traditionally purchased virtually all wool, both quota and over-quota, at procurement prices set by the State.

With the introduction of general economic reforms from 1979 onwards, pressure began to mount for changes to wool marketing. The remaining chapters in this book describe and analyse the implications of the reforms which have been introduced in relation to wool marketing since the mid-1980s. As a result of these changes, major new marketing channels for wool have emerged and the whole agribusiness environment for wool has been transformed. Fig. 2.1 presents a schematic diagram which captures only the broadest elements of the various channels through which wool now reaches Chinese mills. The diagram is very general and is not intended as a definitive statement of the present situation. Furthermore, data are not available on which to base firm estimates of the amounts of wool moving through the various channels in any one year. In any event, the system remains in a state of flux. For example, several of the key State institutions which historically have dominated wool marketing were restructured in April 1993. Consequently, both in this chapter and in subsequent chapters, attention is focused more on the pressures which have led to the institutional changes rather than on a description of the institutions themselves.

2.1.2 Wool Exchange and Service-Related Institutions

Despite the reforms, the SMCs remain the principal buyers of wool at the grass roots, especially in relation to non-State farm wool. The SMCs constitute an enormous agribusiness organisation. An outline of their role and impact on Chinese wool marketing is presented in Chapter 7. Since the mid-1980s, pressure has been mounting for the SMCs to break with the State bureaucracy and re-establish their independence as genuine grower-controlled cooperatives. As discussed in Chapter 7, the SMCs have become more-or-less separate from the State in some wool-growing areas. In general, however, the vast SMC network remains closely linked to the former Ministry of Commerce. In the past, this Ministry controlled not only the SMCs and hence the purchasing of raw wool but also the wholesaling and retailing of wool textiles, yarns and garments. In April 1993, the Ministry of Commerce was reconstituted as the Ministry of Internal Trade. Amongst other things, the new arrangements were intended to accelerate the trend towards a much less regulated market for wool-based products in China.

At the same time, as a result of the deregulation of the domestic trade in raw wool, the SMCs have been challenged both by the growth of direct sales between State farms and mills, and by other government, semi-government and private units wishing to enter the raw wool-marketing business (Sections 5.2 and 7.4). In fact, it is not impossible to imagine that foreign firms may become involved in the marketing of raw wool in China in the near future. For example, it would be relatively easy for foreign buyers from Hong Kong or Taiwan to operate in a revitalised Chinese wool auction system (Chapter 8).

Another institution with a central role in wool exchange in China is the Fibre Inspection Bureau (FIB), which is responsible for wool standards, inspection, testing and grading in relation to both domestic and imported wool. China has developed a remarkable set of formal standards for wool and wool tops, and these are described in

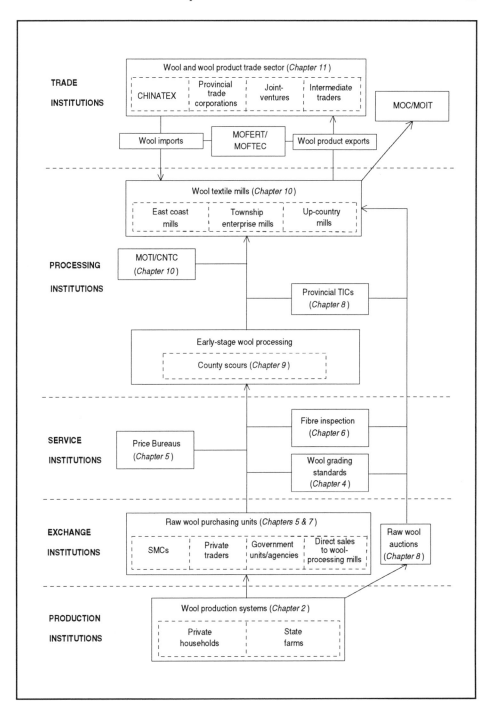

Fig. 2.1 Outline of the Agribusiness Environment for Wool in China

Chapter 4. These standards depend for their effectiveness on the capacity of the FIB to provide adequate fibre-testing and inspection services. Recent reforms have created a much greater need for the services of the FIB, and there are opportunities for foreign fibre-testing R&D agencies to assist the FIB respond to these new pressures (Chapter 6).

In the era of State pricing, the Price Bureaus played a major role in determining prices and in monitoring and policing to ensure that State prices were observed by market participants. Now that State pricing for raw wool and wool-based consumer goods has all but disappeared, the Price Bureaus have taken on a more service-oriented role in relation to the monitoring of prices. In the case of raw wool, there is an urgent need for a public market-reporting service to improve the flow of market information to buyers and sellers. The Price Bureaus are in a good position to fill this important gap but once again, as with the FIB, their capacity to respond to the emerging needs of the wool market is limited without foreign assistance (Chapter 5).

2.1.3 Wool-Processing Institutions

Various institutions have been involved in managing the wool-processing sector. Key among them is the Ministry of Textile Industry (MOTI). An enormous organisation even by Chinese standards, it has had up to 13 million staff involved in the textile industry. MOTI has traditionally controlled the processing of wool tops, wool yarn, knitwear, wool fabrics, wool clothing and wool blankets. It was segmented horizontally into several functional departments of which the most important was the production department which set production goals for mills. Next in importance was a raw material supplies department which obtained raw wool supplies for the mills either through foreign trade organisations or from the SMCs. MOTI was also structured vertically, with ministerial-level units at the top of the organisation which, in theory, coordinated production, investment and raw materials for the entire country. In practice, they issued guidelines and allocated investment funds. At the provincial and local level, individual mills were directly administered by production department units of MOTI. Although these mills were at the bottom of MOTI's structure and were formally State owned, they often had significant autonomy and were usually required to finance the purchases of their own raw materials, including raw wool. Nevertheless, as with other State-owned enterprises, they had access to Central government subsidies should they operate at a loss.

MOTI was another major State organisation affected by the institutional changes of April 1993. Essentially, MOTI was reconstituted as an industry association known as the China National Textile Council (CNTC), responsible directly to the State Council. The change in name and status of MOTI in 1993 was intended to reflect a move away from MOTI's traditional planning and management role towards a greater service and marketing orientation. The changes also implied that the large number of State-owned textile enterprises formerly under the control of MOTI will in future be forced more-or-less to "fend for themselves" without access to Central government subsidies.

These changes in the administrative structure, together with the new pricing arrangements which no longer guarantee returns on a cost-plus basis at each step in the processing chain, are likely to see Chinese wool-processing mills become more discerning buyers of wool. The many large and medium-sized State-owned mills formerly under the guidance of MOTI will face increasing pressures to adjust and become more competitive. Similarly, township-enterprise textile mills are under pressure to perform or perish. In this more competitive environment, mills are likely to continue to seek marketing reforms which enhance both the quality of marketing information and the quality of their raw materials such as wool. A brief outline of the structure of the wool-processing industry in China appears in Section 2.3, but the detailed discussion of the effects of the recent institutional changes and the problems confronting wool-processing mills is left to Chapter 10.

The Ministry of Light Industry coordinates the production of tufted and hand-knotted carpets. However, not all carpet factories are under the control of this Ministry. Carpet factories are also operated by arts and crafts companies, foreign trade companies (Ministry of Foreign Trade and Economic Cooperation) and animal by-product companies within the SMC network (Ministry of Internal Trade). Each of these institutions is responsible for purchasing its own coarse wool used in the manufacturing of carpets. While domestic supplies of coarse wool are important (especially for inland carpet manufacturers), China also imports a large amount of carpet wool from New Zealand and South America.

The Rural Enterprise Bureau under the Ministry of Agriculture monitors the production of rural-township wool mills, all of which fall outside MOTI's jurisdiction. In certain places such as Jiangsu Province, the output of township wool mills is almost as large as that of urban mills in the MOTI/CNTC network. Because township mills have never been subsidised by the Central government, they tend to be more commercially aggressive. Consequently, they react more quickly to market forces and are often more adept than State mills at marketing their products. When the wool boom in China collapsed in 1989 and unwanted output from State mills lay in warehouses all over China, MOTI sought to reduce competition from township mills by restricting their access to raw materials and textile machinery. The conflict during the wool recession of 1989 to 1992 between the MOTI-controlled State-owned mills on the one hand and township-enterprise mills on the other, has had serious consequences for the future of wool processing in China. (See Section 10.1 for more details.)

Since the mid-1980s and especially following the Central government's fiscal reforms, local governments, often through the local SMC, have become involved in early-stage wool processing. This development and some of the unforeseen, adverse effects of it are elaborated in Chapter 9.

2.1.4 Trade-Related Institutions

Imported wool has played a key part in the expansion of the Chinese wool textile industry (Chapter 11). Coastal mills, in particular, have used a high proportion of imported wool in manufacturing products designed for re-export. Changes in wool-marketing arrangements and consumption patterns, however, have led to an interest in importing wool by other mills, including the up-country mills in the wool-growing pastoral region. At various times in the past, the State Planning Commission has set and strictly enforced wool import quotas. Just as there were major changes in domestic marketing arrangements in 1992 and 1993, significant adjustments have also occurred in importing arrangements. In particular, the wool-processing industry is likely to be affected significantly by a downgrading in 1993 of the functions and resources of the Ministry of Foreign Economic Relations and Trade (MOFERT).

Traditionally, MOFERT has had responsibility for raw and semi-processed wool imports and for wool product exports. While the degree of centralised control exercised by MOFERT over the wool trade has varied in recent years, as explained in Chapter 11, this Ministry has been increasingly seen both by Chinese textile mills and by foreign suppliers as a major "bottleneck". A change in its name in 1993 to the Ministry of Foreign Trade and Economic Cooperation (MOFTEC) again reflected attempts to revamp this Ministry.

Wool product exports have been regulated by many of the same institutions that regulate wool imports. Some east coast mills have direct authority to export their textile products, but most mills are forced to operate through the various foreign trade corporations. The Central government through MOFERT has regulated the distribution of export quotas across regions and mills. As with wool imports, joint-venture arrangements have been one approach to circumventing these regulatory constraints on the export of wool-based products. Mills wishing to avoid the foreign trade corporations have operated through those mills with authority to export directly. For instance, a number of the mills in the pastoral region (especially in XUAR and Gansu) have exported through Ren Li mill in Tianjin. Essentially, the Ren Li mill obtains export orders and then subcontracts other mills to produce the products required to fill the orders.

No real wholesaling system exists for textile products in the Chinese domestic market. Individual mills normally employ a large number of travelling salesmen to sell their products directly to department stores and retail outlets in cities throughout China. Retail outlets are typically State-owned under the Ministry of Internal Trade. However, large foreign-owned and -operated retail enterprises have begun to be established in the large cities. For instance, Shanghai now has department stores operated by Johan (Japan), Wungon (Hong Kong), Benetton (Hong Kong) and Stiffile (Italy).

2.2 Wool Production in China

Commercial wool production is a relatively new phenomenon in China, the origins of which more-or-less coincide with the founding of the People's Republic in 1949. The transition from traditional sheep-raising practices to commercial wool-growing systems has been a monumental task and one which is still underway in many parts of the country. Longworth and Williamson (1993) describe and analyse the development of wool growing in China in considerable detail, and the remainder of this section summarises some of the major points made by these authors.

2.2.1 Three Broad Production Systems

The production systems for wool that have emerged in China vary considerably, reflecting both differences in how they have evolved and, in particular, the large variation in environmental conditions under which sheep are raised. Nevertheless the myriad of production systems can be categorised into three broad groups.

Localities in which sheep are raised are classified for administrative purposes according to the relative economic importance of pastoral activities as compared with agricultural activities. That is, a household, village, township, county, prefecture, or province may be classified as pastoral, semi-pastoral/semi-agricultural, or agricultural.

Especially at the household and village level, but also at the township and higher levels in some parts of the pastoral region, the administrative classification also describes the dominant sheep-raising production system. That is, three broad production systems can be identified, depending on the locality in which sheep are raised.

Pastoral production systems dominate in pastoral areas which are located in the semi-desert areas of northern and north western China. Climatic conditions are extreme, with hot dry summers and cold winters. Before 1949, the few people who inhabited these areas were almost all nomadic herders who belonged to minority nationalities. The founding of the People's Republic saw a great increase both in human and livestock populations in these regions. Consequently, almost all herders and their families now live in permanent villages, although livestock are still moved from one grazing area to another on a seasonal basis. In winter (November to April), sheep are grazed near permanent houses during the day and yarded or shedded at night. From May to June and again from late September to early November, sheep are grazed on spring/autumn pastures. From July to September, sheep are grazed on summer pastures which are often remote from the herder's residence.

Semi-pastoral production systems are used in those parts of the pastoral region where rainfall is slightly higher or where irrigation is made possible by the availability of underground water or snow-fed rivers and streams. In these localities, sheep are raised as a major sideline activity to cropping. The number of sheep raised by each household tends to be less than in the purely pastoral areas. While the general availability of forage materials is better, the quality of husbandry is often lower than in the specialist sheep-raising areas.

The agricultural production systems occur in areas located in warmer, moderate rainfall areas where feed availability is better. However, little pasture land is generally available for grazing. In contrast with the pastoral areas, where households tend their own sheep, most sheep in agricultural areas are herded by non-specialist and often hired labour most days of the year. The sheep "scavenge" around the villages, but a substantial component of "cut and carry" feeding is also involved. In agricultural areas, flock sizes are small, with each household usually raising less than five sheep. The quality of sheep husbandry tends to be low, with little attention being given to enhancing wool quality.

2.2.2 Amount, Type and Location of Wool Output

The 240,309 tonne of greasy wool produced in China in 1993 was produced from 112 million sheep. Despite an active sheep-improvement program since the early 1950s (Longworth and Williamson, 1995), some 57% of the national flock are still local coarse-woolled sheep. Sheep breeds are classified according to the type of wool they produce, such as fine wool, semi-fine wool or local/coarse wool. There are a large number of recognised fine wool breeds which are all merino types, including Xinjiang fine wool sheep, Gansu alpine fine wool sheep, Aohan fine wool sheep, North East fine wool sheep and Erdos fine wool sheep. There are also many recognised breeds that produce semi-fine wool, as well as many local breeds or strains which produce coarse or local wool.

Wool, broadly defined, is produced in most provinces and autonomous regions of China. However, in 1993 almost 90% of the sheep were raised, and 83% of the raw wool produced, in the pastoral region. As explained in Section 1.2, the pastoral region is defined as 12 provinces most of which are located in the more remote, less populated north and west regions of China (see Fig. 1.1). The most important provinces in terms of wool growing are the Inner Mongolia Autonomous Region (IMAR), which produced 55,305 tonne or 23% of the greasy wool grown in China in 1993, Xinjiang Uygur Autonomous Region (XUAR), which produced 21%, Shandong 11%, Qinghai 7%, Gansu 6%, and Hebei 5% (Table 2.1). Fine wool and improved fine wool are primarily grown in IMAR, which produced 33,134 tonne or 30% of this type of wool grown in China in 1993, XUAR 24%, Shandong 11%, Jilin 8%, Gansu 6%, Liaoning 4% and Hebei 4%. All of the other provinces or autonomous regions produced less than 2,600 tonne of fine and improved fine wool in 1993.

Table 2.1 Chinese Wool Production by Province and Wool Type, 1993

Province or autonomous region	Fine wool *and* improved fine wool		Semi-fine wool *and* improved semi-fine wool		Total all wool[1]	
	Amount of greasy wool	Proportion of national total	Amount of greasy wool	Proportion of national total	Amount of greasy wool	Proportion of national total·
	(t)	(%)	(t)	(%)	(t)	(%)
Hebei	4,377	4.0	2,722	5.1	11,182	4.7
Shanxi	2,461	2.2	689	1.3	6,170	2.6
IMAR	33,134	30.1	12,198	22.7	55,305	23.0
Liaoning	4,219	3.8	2,410	4.5	7,370	3.1
Jilin	8,573	7.8	1,077	2.0	9,869	4.1
Heilongjiang	3,972	3.6	7,020	13.1	11,016	4.6
Sichuan	276	0.3	614	1.1	3,040	1.3
Tibet	33	0.0	2,681	5.0	8,376	3.5
Gansu	6,470	5.9	1,164	2.2	15,411	6.4
Qinghai	305	0.3	6,074	11.3	17,044	7.1
Ningxia	265	0.2	--	--	3,021	1.3
XUAR	26,036	23.7	5,774	10.8	50,818	21.1
Sub-total of 12 pastoral provinces	90,121	82.0	42,423	79.1	198,622	82.7
Shandong	11,916	10.8	6,610	12.3	25,746	10.7
GRAND TOTAL FOR ALL CHINA	109,969	100.0	53,634	100.0	240,309	100.0

--, Negligible.
[1]In addition to fine and semi-fine wool, there is a considerable amount of coarse and local wool (primarily used for making carpets, upholstery, tents and felt) grown in China. In 1993 the production of this "other" wool was 76,706t.
Source: China Agricultural Yearbook (1994).

2.2.3 Organisation of Production

Private households now own almost 90% of all sheep in China. The private sheep farmer operates under the household production responsibility system which was introduced throughout rural China after 1978. In the pastoral region, the household production responsibility system was usually implemented in three steps. First, the livestock previously owned by communes were assigned to households on a rental basis. Second, the livestock were sold to households. Third, the land owned by the collectives under the commune was contracted out to private households. Consequently, private households now have ownership of their livestock, although they have responsibilities to their local (village) collectives both in the form of welfare taxes and in their use of specified areas of pasture.

The remainder of the sheep in China are raised on State farms. Although many of these State farms have introduced a kind of household production responsibility system which gives the individual households on these farms responsibility for the day-to-day management of the sheep, the administration of the State farm retains overall control of the flocks. As a result, the State-farm flocks are usually much larger and much more homogeneous with respect to wool characteristics than are the flocks of the private sheep herders.

State farms play a major part in the Chinese domestic wool-growing industry. Although they produce little more than 10% of the total Chinese greasy-wool clip, most of this wool is in the fine and improved fine category (i.e. has a mean fibre diameter of ≤25μm—see Chapter 4 for more details on the definition of fine wool). Longworth and Williamson (1993, p.65) estimate that about one-fifth of all fine and improved fine wool grown in China is produced on State farms. Furthermore, the proportion of sheep which are genuine fine wools (rather than crossbred or "improved" fine wools) would be much higher than is the case for privately-owned sheep. Therefore, perhaps about half the *genuine* fine wool (as compared with *improved* fine wool) grown in China is produced on State farms.

Almost all wool-growing State farms belong to one of two separate State farm networks. While both networks are ultimately under the Ministry of Agriculture, they are controlled by different departments within the Ministry and they compete vigorously for resources, especially foreign aid. The Production and Construction Corps (PCC) farms under the State Farms and Land Reclamation Department are especially significant in XUAR. The other major group of wool-growing State farms are controlled by the Animal Husbandry and Veterinary (Science) Department and are commonly referred to as the Animal Husbandry Bureau (AHB) network. For a number of reasons, as explained by Longworth and Williamson (1993, pp.65–67), it is important for foreigners to be aware of the difference between PCC State farms and State farms in the AHB network.

2.2.4 Wool Quality

Compared with wool of similar fibre diameter produced elsewhere in the world, Chinese fine and improved fine wool suffers from a number of quality problems. Lack of fibre strength or tenderness is especially important, although the extent of this problem varies from year to year. The sheep are usually shorn in early summer (June/July). As a result, each fibre in the staple tends to be thicker towards the tip because it grows while the sheep are enjoying better nutritional conditions in summer, whereas the bottom of the fibre is weaker and thinner because it grows in winter and in early spring (November to April inclusive). There is often at least one distinct break or weak point in the fibre corresponding with a period during the year when the sheep suffered severe physiological stress due to illness, extremes of climate or a sudden change in diet. Since these physiological shocks often occur with the onset of winter (November), the break in the wool is frequently near the middle of the fibre. An associated problem is that the thick end of the fibre at the top of the staple or tip is commonly damaged by ultraviolet light, especially if the wool is grown on sheep raised in alpine districts. Another general issue frequently raised by mill managers is that a lot of domestic wool is too short. For example, typically fine and improved fine wool grown in the IMAR has an average staple length well below the 8cm considered necessary for the production of good yarn. Wool contamination is another major problem which varies in severity from district to district and from year to year. Clean yields are particularly low by international standards, averaging around 35% in the major wool-producing areas of the IMAR. Average clean yields for fine wool in XUAR are higher at just over 40%. By way of comparison, the average yield of clean wool for the whole Australian clip is about 65%.

A key factor limiting the future potential of the Chinese wool-growing industry to improve both the quality and quantity of its output is the extent to which pastures are deteriorating. The rangelands of the pastoral region are already severely degraded, with clear evidence of a trend to even further deterioration. Liu (1993) states that degraded grasslands in the pastoral region amount to 86.7 million ha or around one-third of the total useable grassland. Liu also argues that rangeland degradation is accelerating. Zhou (1990) cited evidence that the degradation process had resulted in a fall in average rangeland dry-matter production from 1.91t per ha per annum in the 1950s to 1.29t per ha in the 1980s. By the end of the 1980s, the annual loss of rangeland in China due to degradation was estimated at 0.7 million ha (Anon, 1989).

2.2.5 Political and Strategic Significance of Wool Production

Traditionally, sheep raising in China was conducted by ethnic minorities such as the Mongolians, Tibetans, Kazaks, Uygur, Hui, etc. These groups made up the majority of the population in the remote, sparsely populated pastoral areas until the founding of the People's Republic in 1949. Despite the great influx of Han Chinese into these regions since that time, especially in the 1950s and 1960s, minority nationalities still predominate in the western provinces, accounting for 62%, 96.5% and 55.6% of the total population in XUAR, Tibet and Qinghai respectively (*China Statistical Yearbook*, 1990, p.76). The minority populations wield significant decision-making power in these and other parts of the pastoral region. For instance, many of the government and administrative units (provinces, prefectures and counties) in the pastoral region are referred to as autonomous units. These autonomous administrative units are allowed a limited amount of self-government consistent with the Constitution and State Laws. Special conditions also seek to provide a "sufficient" level of national representation, and priority is supposed to be attached to accelerating development in autonomous areas (Longworth and Williamson, 1993, pp.37–42).

Despite the longstanding official emphasis on improving economic conditions in the pastoral region, income differences between the west, central and east regions of

China widened significantly (perhaps by 30%) between 1979 and 1992. There are now major differences between the level of economic development in these three broad regional groupings. For example, average rural per capita net income for 1992 in the nine provinces of the west was ¥600, for the 10 provinces included in the central region it was ¥900, while the corresponding statistic for the 11 eastern provinces was ¥1,300 or more than double the figure for the west region (ARGRE, 1994, pp.125–129). Three broad explanations are usually given for why investment opportunities and growth have been greater in the east. First and most importantly, infrastructure and resource endowments (especially human capital) favour the east. Second, the east and to some degree the central regions have a history of greater market orientation and commercialisation. Third, the reforms penetrated further down the government hierarchy faster in the east, thus giving the east "a start" on the rest of the country.

With widespread access to television and other media, the inhabitants of the west region are becoming increasingly aware of the widening gap between their circumstances and conditions elsewhere in China. While it is easy to exaggerate the extent of the problem, civil unrest, motivated in part by the lack of economic opportunities, has from time to time been a reality in the western provinces. For example, official concern about the risk of major political disturbances forced the XUAR Textile Industry Corporation to transfer its wool auction planned for Urumqi in mid-1991 to Xi'an (Section 8.3.1). The Central government, therefore, is becoming increasingly conscious of the need to encourage economic development in the politically and strategically sensitive pastoral region (Longworth and Williamson, 1993, pp.41–42).

One reason for the income disparity is the fewer off-farm opportunities in pastoral areas. Traditional agricultural and animal husbandry activities such as wool growing still generate the bulk of rural income (87% in XUAR, 77% in IMAR, 94% in Tibet, and 81% in Qinghai). For a variety of reasons, but principally owing to remoteness from product and input markets, rural-township enterprises have not flourished in the pastoral provinces in the same way that they have in the eastern provinces such as Jiangsu and Shandong.

Given the limited resource endowments in pastoral areas, it is not surprising that animal husbandry-based industries such as wool processing are being given renewed emphasis by the Central government. For political rather than economic reasons, it is essential that a viable wool textile industry continues to exist in the main wool-growing provinces of north and west China.

2.2.6 Implications for Wool-Marketing Reform

The characteristics of wool growing and the special features of the pastoral region affect China's wool-marketing system in a number of important ways. To be successful, reforms need both to address the problems and to recognise the constraints created by the physical, economic and socio-political environment in which wool is grown.

Raw wool purchase: The wool purchase system needs to accommodate the wide geographic dispersion in production and the heterogeneous nature of the wool. As mentioned earlier, almost 90% of wool is now produced by an enormous number of individual herders who produce small quantities of many different types of wool. This has implications for the aggregation and segregation of wool lots. It also means that there must be a great number of individual purchase stations which are small and remote, militating against grass-roots purchase systems that require high levels of grading skills and complicated equipment.

Wool grading: Paradoxically, some of the reasons which inhibit the application of advanced purchase systems only serve to emphasise the need to develop an appropriate system of grading. That is, the large number of dispersed wool growers living in remote areas means that the point of production is often far removed from the point of processing. Under these conditions, it is impractical for mills to buy directly from growers. Hence, there is a need for intermediaries to perform the assembly function. Such marketing intermediaries need to be guided by a grading system which reflects the requirements of the mills but which at the same time is an appropriate means of sorting the wool at the time it is purchased from the producers. An effective grading scheme of this type is needed so that mills are fully aware of what they are buying from the intermediaries and so that the preferences of mills for different types of wool can be transmitted to individual herders.

Wool contamination: The extremely low clean yields obtained from raw wool in China have major implications for marketing and processing. Both the nature and the extent of the contamination need to be taken into account when the wool is purchased from the grower. This implies the need for a reasonably sophisticated grading system and a means of measuring (or at least estimating) clean yields. Traditionally, purchasing has been on a raw or greasy basis without any regard to the nature of the contamination. Clearly, major reforms are necessary in this aspect of the marketing system.

Another important consideration is the effect heavy contamination has on first-stage processing. It is much more difficult to scour heavily soiled wool without damaging the fibre. As wool quality becomes more important in China, much greater attention is likely to be given to the question of wool contamination and its implications for scouring.

Staged introduction of reforms: Many of the wool-marketing reforms over the last decade have been trialled on State farms. State farms are the only wool-growing entities with sufficient homogeneous supplies of the better-quality wool to adequately test the new marketing arrangements. Once these reforms have been accepted in the State-farm sector, it will still be necessary to adapt the new arrangements to the needs of the private-household sector which collectively grows most of the wool but which individually produces very small quantities of heterogeneous wool.

Regional autonomy: Much of the wool is grown in regions in which ethnic minorities make up the majority of the population. Implementation of wool-marketing reforms initiated at the central level is influenced by the degree of local autonomy in these regions. In some cases, local autonomy means that wool marketing is not well coordinated, and parochial inter-regional competition can and does arise.

Value-adding activities: The geographic remoteness, limited industrial opportunities and relatively low incomes of the communities in which wool is produced have led to the establishment of value-adding activities related to animal husbandry such as local wool scours and topmaking facilities. Unfortunately, however, owing to the parochial inter-regional competition mentioned above, the development of these facilities has often had disastrous consequences.

Wool imports: The severely degraded state of the rangelands limits future livestock numbers and constrains the ability of the Chinese wool textile industry to expand on the basis of local wool supplies. Although new technology may improve the efficiency of feed conversion as well as enhance wool quality, it is unlikely to dramatically increase the amount of local wool available to Chinese mills. Consequently, future marketing arrangements need to accommodate greater quantities of imported wool.

Political and strategic issues: The increasing political instability in Mongolia, Russia, Kazakhstan, Kirghizstan, Tadzhikistan, Afghanistan, Pakistan and northern India (Kashmir), all of which have common borders with China's pastoral region, emphasises the strategic sensitivity of the pastoral provinces. Wool growing and wool processing are important economic activities in northern and north western China. Furthermore, many of the minority-nationality people living in this part of the country, but with ethnic links across the border, depend upon sheep and wool for their livelihood. Access to imported wool may be essential if the up-country wool mills in the pastoral region are to remain economically viable in the longer term. However, it is also imperative that these mills continue to purchase locally grown wool from the traditional pastoralists. The need to protect the incomes of the wool-growing minorities in remote but strategically critical parts of the country has emerged as a major constraint to further reform of China's wool-importing arrangements. Indeed, wool import policy is proving a serious problem for China in its attempts to rejoin the GATT and to become a founding member of the proposed World Trade Organisation.

2.3 Wool Processing in China

Wool processing is a multi-stage operation. The early stages (e.g. scouring and topmaking) are sometimes undertaken close to where the raw wool is grown, while the more advanced processing (e.g. woollen and worsted manufacturing) is usually carried out in large mills located in industrial centres. Prior to 1978, most raw wool grown in China was processed by large State-owned integrated mills located either in major inland centres such as Lanzhou, the capital of Gansu Province, or in big eastern cities such as Beijing, Tianjin, Nanjing, and Shanghai. The post-1978 economic reforms stimulated investment in wool processing. The capacity of the industry as measured by the number of wool spindles and looms expanded remarkably (Table 2.2). The total number of wool spindles in China rose from 478,100 in 1978 to 3.3 million in 1992, while the number of looms increased from 7,120 to 35,131 over the same period. The large State enterprises expanded and modernised their equipment, but the really big jump in processing capacity came from the establishment of a large number of new rural-township-enterprise mills from about 1983 onwards. These township enterprises were mainly concentrated in the eastern provinces, especially Jiangsu. In addition, in the second half of the 1980s, many remote wool-growing counties established early-stage processing plants.

Table 2.2 Wool-Processing Capacity of China, 1978 to 1992

Year	Wool spindles				Wool looms
	National total	Yarn	Worsted	Woollen	
	- - - - - - - - - - - - - - - - ('000) - - - - - - - - - - - - - -				(no.)
1978	478.1	95.8	268.5	94.8	7,120
1979	532.9	111.3	292.9	107.5	7,742
1980	600.5	176.1	295.1	103.3	8,332
1981	744.4	228.6	328.0	139.5	10,054
1982	888.8	375.4	300.7	166.4	12,446
1983	1,005.3	308.5	445.1	209.5	14,650
1984	1,205.2	325.1	497.6	283.7	17,121
1985	1,394.9	368.6	606.6	379.5	21,676
1986	1,685.3	525.3	731.9	392.8	25,704
1987	1,992.0	638.6	821.0	489.3	29,171
1988	2,266.7	756.1	898.5	562.2	33,451
1989	2,522.7	881.5	969.2	598.1	32,602
1990	2,658.7	941.4	1,010.2	638.0	33,556
1991	3,030.2	1,110.7	1,145.8	709.8	35,128
1992	3,292.5	1,292.3	1,187.9	736.2	35,131

Source: *Almanac of Chinese Textile Industry* (various issues).

Wool processing in China was completely transformed during the 1980s. Not only was there a massive increase in overall processing capacity but there were also major changes in the structure and location of the industry and in the composition of its output. Zhang (1990a), Findlay and Li (1992), and others have reported the developments in the Chinese wool textile industry in detail. This section draws on and extends these earlier reviews to highlight the implications of some of these changes for the Chinese wool-marketing system.

2.3.1 Changes in the Pattern of Output

While in absolute terms the expansion in wool processing since around 1980 has been spectacular, the growth in this industry between 1950 and 1980 was also remarkable. Zhang (1990a) provides detailed statistics, but the essence of the story is captured in Fig. 2.2 which shows the growth in wool textile output since 1950 in terms of three major product categories: yarn, fabric and blankets. In 1950, only 1.3kt of wool yarn, 4.88 million metres of fabric and 485,000 blankets were produced. Output expanded greatly from these low bases in the 1950s, 1960s and 1970s, with yarn production increasing more than 40 times and fabric and blanket production almost 20 times by 1980. Since the late 1970s, there has been even faster growth in wool textile production in most years. As was the case in the earlier decades, the largest growth occurred in wool yarns, with an average annual growth rate of around 16%. The growth rates for fabrics and blankets were around 11% per annum. Despite these high average annual growth rates, the wool textile industry experienced a major boom/bust situation during the 1980s. For instance, production of blankets almost doubled between 1984 and 1988 but fell away sharply in 1989 and 1990.

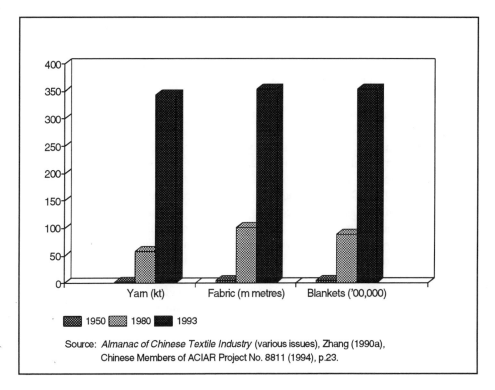

Source: *Almanac of Chinese Textile Industry* (various issues), Zhang (1990a),
Chinese Members of ACIAR Project No. 8811 (1994), p.23.

Fig. 2.2 Chinese Wool Textile Production by Type, 1950 to 1993

Within the wool yarn sector, much of the expansion has been in fine yarn for use in machine knitting. Worsted yarn for knitting now accounts for half the wool yarn produced in China. This sector is one of the most capital-intensive and the one experiencing the fastest growth in exports (Findlay and Li, 1992). Wool fabric production is equally divided between worsted fabrics and woollen fabrics. The developments in these two sub-sectors, however, have been quite different. Having commenced from a similar low production base in 1950, worsted fabric production exceeded woollen fabric production by a factor of three at the start of the 1980s. Specifically, worsted fabric production rose from 1.55 million metres in 1950 to 67.48 million metres by 1980, while woollen fabrics increased from 2.91 to 23.32 million metres over the corresponding period. However, this trend was subsequently reversed following the post-1978 economic reforms. That is, during the 1980s worsted fabric production doubled while woollen fabric production increased over six times. The rapid rise in woollen fabric production was associated primarily with the emergence of a large number of township enterprises processing wool in the eastern provinces.

The changing pattern of output reflected the adjustments in spindle capacity. Of the 129,000 wool spindles in 1949, half were worsted spindles, one-third woollen spindles, and one-sixth knitting wool spindles. By 1980, while about half the then 600,500 spindles were still in the worsted sector, the woollen sector accounted for only about one-sixth of the total industry capacity. There had been a big shift towards yarn spindles which represented almost 30% of all spindles in that year (Table 2.2). The relative share of yarn spindles further increased during the 1980s and early 1990s to reach roughly 40% of the total 3.3 million spindles in 1992. During this period, the woollen sector also expanded faster than the wool textile industry as a whole. On the other hand, while the number of worsted spindles increased more than three-fold between 1980 and 1992 (Table 2.2), the worsted sector share of the total spindles declined from around 50% to less than 37% over this period.

2.3.2 Changes in the Location of the Industry

As might be expected, given the major expansion in wool textile processing capacity since 1950, there has also been a substantial change in the location of the industry.

Table 2.3 presents the distribution of spindles by provincial-level administrative units in 1952, 1970, 1980, 1985 and 1990. Two features of these data stand out. First, wool textile manufacturing has become a widespread industry in China. Twenty-four of the 29 provinces/cities listed in Table 2.3 had more than 25,000 wool spindles in 1990. There were only seven provinces/cities with this number of wool spindles in 1980 and only one in 1952. The second major feature in Table 2.3 is the rise of Jiangsu as a wool-processing centre. In 1952, there were only 3,000 wool spindles in Jiangsu. By 1980, this had risen to 82,000, which made it second to Shanghai in importance as a wool-processing locality at that time. But by 1990, Jiangsu had more than 644,000 spindles, which was almost three times as many as Shanghai had.

Another important development illustrated by the data in Table 2.3 is the expansion in the wool-processing capacity of the major wool-growing provinces, especially XUAR, IMAR and Gansu. In 1952, IMAR and Gansu had a total of just over 2,000 spindles and XUAR had none. By 1980, all three of these provinces had a significant wool textile manufacturing capability, with a total of 75,000 wool spindles between them. Ten years later, in 1990, XUAR had 98,600 spindles, Gansu 74,500 and IMAR 71,900. During the 1980s, the combined wool textile capacity of these three up-country provinces grew more rapidly than that of most other parts of China, with the notable exception of Jiangsu.

2.3.3 Emergence of Three Sectors

There are now wool-processing establishments in virtually every province in China (Table 2.3). These mills can be classified into three potentially useful though artificial groups (Zhang, 1990a). The first group are the large integrated mills owned by the State and located in the eastern provinces of Liaoning, Beijing-shi, Tianjin-shi, Jiangsu (Nanjing), Shanghai-shi and Zhejiang. The second group of mills are the rural-township enterprises which emerged following the economic reforms in the 1980s. These mills are primarily located in the eastern coastal provinces especially Jiangsu but also Shandong, Zhejiang and Guangdong. Up-country mills, especially the mills located in the wool-producing pastoral provinces of northern and north western China, make up the third group of mills.

Table 2.3 Distribution of Wool Spindles by Province: 1952, 1970, 1980, 1985 and 1990

Province (city)	1952	1970	1980	1985	1990
			('000 spindles)		
Beijing	10.2	37.4	58.6	74.1	92.8
Tianjin	15.9	33.7	36.6	92.4	95.4
Hebei		0.7	15.6	31.0	81.1
Shanxi	1.0	7.7	13.0	19.6	28.5
IMAR	0.8	16.7	26.4	41.0	71.9
Liaoning	18.7	26.8	43.0	77.9	119.5
Jilin		3.7	11.6	35.3	48.4
Heilongjiang	1.7	10.3	15.1	41.0	86.0
Shanghai	77.2	103.5	125.1	176.4	215.0
Jiangsu	3.0	16.7	82.0	270.4	644.7
Zhejiang		3.9	15.7	50.8	154.1
Anhui		1.3	3.0	23.8	62.6
Fujian			1.9	12.6	29.2
Jiangxi			1.0	12.4	28.8
Shandong			18.6	92.8	218.5
Henan		1.1	16.4	53.3	92.7
Hubei			9.4	45.7	108.2
Hunan			5.8	16.3	27.1
Guangdong			10.4	29.7	74.2
Guangxi				4.6	20.5
Sichuan	7.4	9.1	9.2	16.4	55.1
Guizhou				5.2	8.1
Yunnan		0.5	1.8	8.8	12.5
Xizang		0.8	2.2	2.3	2.3
Shaanxi	0.1	8.0	13.8	27.8	51.6
Gansu	1.3	12.2	31.2	54.4	74.5
Qinghai		6.1	10.4	20.2	35.2
Ningxia		0.5	5.6	13.0	21.5
XUAR		13.1	17.4	45.7	98.6

Source: *Almanac of Chinese Textile Industry* (various issues).

Until the 1980s, wool processing in China was primarily identified with the first group of mills. Indeed as already pointed out in relation to the discussion of the data in Table 2.3, virtually all the wool spindles in China 40 years ago were located in the cities of Shanghai, Tianjin, Beijing and in Liaoning Province. While the industry was much more geographically dispersed in 1980, Table 2.3 again demonstrates that the traditional east coast centres still dominated the industry. Foreigners interested in wool processing in China traditionally focused their attention on this group of mills. Under MOTI guidance and control, these mills generally have had access to the best equipment and wool supplies (both domestic and imported). Many of the mills seek to produce higher-quality fabrics and have a disproportionately large share of Chinese worsted fabric production (Table 2.4). Although Shanghai has since lost its absolute dominance of the wool textile industry, there was a significant increase in processing capacity (especially worsted spindles) in Shanghai mills prior to 1980.

Table 2.4 Characteristics of Wool Processing in Selected Provinces and Selected Years

	Wool spindles				Proportion of Chinese spindles (1990)	Proportion of Chinese worsted production (1990)	Proportion of Chinese woollen production (1990)	Proportion of Chinese blanket production (1990)
	1952	1970	1980	1990				
	- - - - - - - - ('000) - - - - - - - -				- - - - - - - - - - - - - (%) - - - - - - - - - - - - -			
Beijing/ Tianjin	26.1	71.1	95.2	188.2	7.1	13.8	7.7	6.4
Hubei			9.4	108.2	4.1	2.3	1.6	5.6
Jiangsu	3.0	16.7	82.0	644.7	24.2	15.3	35.5	10.8
Liaoning	18.7	26.8	43.0	119.5	4.5	6.0	5.7	4.8
Shandong			18.6	218.5	8.2	6.8	6.0	11.8
Shanghai	77.2	103.5	125.1	215.0	8.1	15.4	6.3	6.6
Zhejiang		3.9	15.7	154.1	5.8	4.3	12.2	3.9

Source: *Almanac of Chinese Textile Industry* (various issues), Zhang (1990a).

The generally higher quality of the products of these big eastern mills compared with other Chinese mills has meant they have been more export-oriented. They have also tended to use a high proportion of imported wool in their overall production. For instance, 90% of the wool used by the Beijing No. 1 Worsted Mill is imported. Altogether, 50% of the mill's output is directly exported to over 50 countries, while some of the remaining output supplied to the domestic market is purchased by garment manufacturers who also export their products. Most of the relatively small proportion of domestic wool acquired by the east coast mills is normally the better-quality wool (special grade fine wool) produced in select parts of IMAR and XUAR. Despite their typically better-quality raw wool supplies and equipment, these mills face the same inefficiencies and are experiencing the same pressures that are confronting other large State-owned enterprises in China.

The expansion in the Chinese wool textile industry during the 1980s was largely the result of the emergence of the second group of mills. The explosion in rural-township enterprises (many of which were involved in textile and garment manufacturing) was a key element of the economic growth in China during the 1980s. The post-1978 reforms enabled local governments to establish textile factories outside the centralised development plans administered by MOTI. Much of the expansion in wool textile manufacturing capacity was centred in Jiangsu Province. Although Jiangsu had become an important wool-processing centre in China by 1980 (accounting for around one-seventh of Chinese processing capacity), spindle capacity exploded in the 1980s, rising from 82,000 spindles at the start of the decade to 644,700 spindles by 1990 or almost one-quarter of Chinese wool-processing capacity. Table 2.4 also reveals how other eastern provinces such as Shandong and Hubei with virtually no history in wool processing prior to 1980 became significant wool-processing provinces during the 1980s. Initially, many of these mills commenced with second-hand machinery purchased from the large State-owned mills which were renovating at that time under the Sixth Five-Year Plan. The township-enterprise mills typically focused on woollen fabric production or, in the case of Shandong and Hubei, on blanket production (Table 2.4).

The rapid growth and entrepreneurial activity associated with many of these township-enterprise mills have tended to focus much of the recent foreign interest in the Chinese wool industry on these mills. To gain access to foreign technology and wool supplies, many of these enterprises have entered into joint-venture arrangements with foreign companies. Joint ventures are becoming increasingly prevalent and increasingly important in the Chinese wool industry. Garnaut *et al.* (1993) note over 60 joint ventures in wool operations in China, involving Hong Kong, Japan, Taiwan, Western Europe (especially Germany), Korea and Australia. As discussed in some detail in several later chapters of this book, joint-venture arrangements have been used for a variety of purposes, including access to equipment, technology, finance and market information, as well as for bypassing restrictive import and export arrangements. Barring policy reversals regarding joint ventures, such arrangements are likely to develop further within the Chinese wool textile industry and to expand to other areas of the wool-marketing chain.

The third group of mills are the up-country mills in the pastoral region. These mills are closely linked with the Chinese wool-growing industry. Although these mills account for a little less than one-tenth of wool-processing capacity in China, they are of special importance in the context of this book because of their links with domestic wool production and wool-marketing arrangements. Furthermore, unlike the situation in relation to the other two groups of mills, relatively little is known outside China about the up-country mills. Consequently, some further details about the up-country mills are provided in the following section.

2.3.4 Wool Textile Industry in the Main Wool-Growing Provinces

As explained in Section 2.2.2, over half the raw wool grown in China is produced in XUAR, IMAR and Gansu Province. Even more importantly in relation to wool textile manufacturing, over two-thirds of the fine and improved fine wool comes from these three provinces. Wool mills in these three provinces constitute an overwhelmingly large proportion of the up-county sector of the wool textile industry but represent only 9.2% of total wool-spinning capacity in the country (Table 2.5). Many of the mills are large, State-owned enterprises with a long history of wool processing sometimes extending back to the 1920s and 1930s. The Gansu No. 1 and No. 3 mills located in Lanzhou, the provincial capital of Gansu, are two of the largest wool-processing mills in China, with 17,000 and 20,000 spindles respectively and with a history extending back more than 40 years. Similarly, the IMAR No. 1 mill in the capital, Huhehot, was established in 1962 and has 15,000 spindles. Apart from the large mills, there are a number of medium-sized mills in regional centres, especially in the IMAR and XUAR. All these mills traditionally belonged to the State-owned MOTI network, although they were seen as the poor cousins of the east coast mills. In the latter half of the 1980s, this processing capacity was extended by local authorities establishing a large number of small wool mills many of which subsequently closed or were mothballed during the wool market crisis at the end of the decade.

For much of their history, the up-country mills have been largely restricted to using wool produced in their region and have had only limited access to imported wool. Wool-marketing reforms in 1985 included a national policy of "self-produce, self-process, self-sell" for the four western provinces of XUAR, Qinghai, Gansu and IMAR. Under this policy, described in detail in Section 3.3, the movement of raw wool between provinces was severely curtailed. Pastoral-region mills received almost no import quota allocations after 1985. Further details about wool imports in the pastoral region are discussed in Section 11.2. Some key aspects of wool processing and production in XUAR, IMAR and Gansu Province are presented in Table 2.5.

The XUAR produces a little more than one-fifth of China's greasy wool or 50kt per annum. About 35kt of this wool can be used by textile mills, while the remainder is coarse/local wool used for carpets, upholstery, tents and felt. Over 200 State farms form an integral part of the wool production system in XUAR, as they typically raise most of the better-quality sheep and produce some of the best wool grown in China. In 1992, the 36 wool-processing mills in XUAR accounted for around 100,000 spindles (65,000 worsted spindles, 18,000 woollen spindles and 17,000 yarn spindles). Two distinct types of mills exist in the XUAR. First, there are the large or medium-sized textile mills located in the main wool-producing areas, for example, XUAR No. 1 mill at Changji with 11,300 worsted spindles; Yili mill in Yining City with 15,000 worsted spindles; August 1 mill at Shihezi with 15,000 worsted spindles; and other medium-sized mills at Aksu City and Urumqi. Second, since the mid-1980s, fully integrated woollen mills with an average size of only 300 to 400 spindles have developed at the county level.

In 1992, more than 13 million metres of fabrics (worsted and woollen), 4,000 tonne of yarn and 300,000 blankets were produced in the XUAR. Like other inland regions, the XUAR produces a significantly higher proportion of worsted fabrics relative to the national average (Table 2.5). In comparison with IMAR and Gansu province, however, a much higher proportion of the wool fabrics produced in the XUAR are pure-wool rather than wool/synthetic blends.

The IMAR is the largest wool-producing region in China, with around one-quarter of China's total wool production. The total number of wool spindles in 1991 amounted to 71,900. The major mills are located in the main producing regions as well as in the capital, Huhehot. For instance, the mills in Huhehot include the IMAR No. 2 mill with 15,000 spindles producing pure-wool and blended fabrics, the IMAR No. 1 woollen mill with 7,188 spindles, and a large, specialist topmaking mill. Both the IMAR No. 1 and No. 2 mills are IWS Woolmark licensed mills. Apart from the wool mills, Huhehot has two carpet-making factories belonging to the Ministry of Light Industry and seven clothing factories belonging to the Light Industry Bureau. Large or medium-sized mills are also located elsewhere in the province, such as at Chifeng City and Tong Liao in the Eastern Grassland wool-growing area of IMAR. Worsted fabrics are again the major fabrics produced at these mills, although wool textile mills in IMAR use more synthetics and produce more blended products than those in XUAR (Table 2.5).

Table 2.5   Overview of the Wool Textile Industry in Selected Major Wool-Growing Provinces of China, 1990[1]

	Xinjiang Uygur Autonomous Region (XUAR)	Inner Mongolia Autonomous Region (IMAR)	Gansu Province	People's Republic of China
Raw wool production ('000t greasy)				239.6
– % of national Chinese total	20.6	24.9	6.3	100.0
Fine wool ('000t greasy)				108.6
– % of national total	26.3	30.1	4.1	100.0
Wool spindles ('000)				2,658.7
– % of national total	3.7	2.7	2.8	100.0
Wool fabric (million metres)[2]				295.1
– % of national total	3.5	3.9	3.0	100.0
(a) Fabric type				
– % worsted	71.0	63.2	70.0	48.5
– % woollen	29.0	35.9	30.0	50.6
(b) Raw material type				
– % pure wool	89.6	52.9	24.9	30.9
– % blended	10.4	47.1	75.1	69.0
Wool yarn ('000t)				249.9
– % of national total	1.4	2.7	2.9	100.0
(a) Yarn type				
– % coarse	58.8	71.6	68.0	43.0
– % fine	0.0	4.9	5.2	7.6
– % worsted knit	41.2	23.5	26.8	49.4
Wool blankets (million pieces)				23.0
– % of national total	1.4	2.5	8.4	100.0

[1] Provincial level information on wool fabrics, wool yarn and wool blankets was derived from Zhang (1990a) and so based on 1988 data.
[2] Percentages for fabric type (worsted, woollen) do not include the Chinese categories of plush and lambsdown and so do not add up to 100 in all cases.
Source: *Almanac of Chinese Textile Industry* (1992), *Agricultural Yearbook of China* (1992), Zhang (1990a).

Gansu Province produces much less raw wool than either the XUAR or IMAR. Nevertheless, as mentioned previously, it has a long history of wool processing, with some of the largest and oldest wool textile mills in the country. Of the 80,000 spindles in the province, 30,000 are for worsted production and 40,000 for woollen production, including blankets. Yarn output, both woollen and worsted and including synthetics, is around 6kt per annum. Most of the wool is processed at two large mills in the capital, Lanzhou. One of these, Gansu No. 1 mill, has 10,864 worsted spindles and 5,520 woollen spindles and produces around 5.6 million metres of fabric, of which around one-third is exported to Hong Kong, South East Asia and the Middle East. Altogether, the mill makes around 200 products which can be further divided into 400 types. Since 1988, the IWS has registered 10 products of the mill with the Woolmark symbol and this may entail significant price and market volume advantages for the mill. The other major mill in Lanzhou, Gansu No. 3 mill, has around 20,000 spindles. A large yarn mill producing around 5,000 tonne of yarn per year also operates at Tianshu City. Gansu mills use a much higher proportion of synthetics in their wool fabrics than either XUAR or IMAR, reflecting, among other things, a larger production of chemical fibres in Gansu Province. Total derived demand by Gansu mills for raw wool is about 5kt (clean-wool equivalent) per annum, half of which is met by wool from the province and half by wool from other provinces and by imported wool.

2.3.5 Implications for Wool-Marketing Reform

The special features of Chinese wool processing affect China's wool-marketing system in a number of ways. Some of the characteristics of wool growing discussed in Section 2.2.6, such as regional autonomy, the need for value-adding activities, and the political and strategic importance of wool are also relevant in relation to processing, especially for that processing which occurs in the pastoral region. However, there are other unique characteristics related to wool processing which should be noted for their influence on wool marketing.

Remoteness from producing regions: Despite the growth in processing capacity in the pastoral region, the majority of mills which rely on domestic supplies of wool are located well away from the pastoral region in the eastern parts of China. Even the "up-country" mills tend to be located in the provincial or prefectural capitals often at some distance from where the wool is grown. When this geographic separation is combined with the dispersion of production within the pastoral region and the typically small lots produced by each production unit, then it becomes extremely difficult for the mills to identify and locate the wool that they require. Thus the marketing system has a major role to play in facilitating the exchange of wool between production and processing regions.

Rapid expansion of wool processing: The rapid expansion in Chinese wool processing since the economic reforms of the late 1970s has outstripped the marketing infrastructure needed to support such a rapid development. Traditional grading systems and information networks may have been adequate in the past under the centralised and regulated marketing systems but these old systems have been unable to cope with the rapid growth, increasing commercialisation and decentralisation of the industry. Major changes in many aspects of marketing have been and are still required to satisfy the requirements of the more sophisticated and much expanded wool textile industry.

Diversity in Chinese wool processing: The three sectors of the Chinese wool-processing sector identified in Section 2.3.3 vary greatly in the source of their supplies, their market orientation, their technical capability, the problems they face and their special needs. The rapid development that has occurred within the wool-processing industry has increased the diversity in equipment and skill levels across mills. The challenge for the Chinese wool industry is to develop marketing systems which can accommodate the diverse requirements of the mills while at the same time encouraging an overall improvement in mill efficiency.

Pressures to modernise and improve product quality: As China increases its involvement in world textile trade and competes more directly with other wool textile-producing countries; as the incomes of its own consumers improve and they demand better-quality textiles and garments; and as wool faces increasing competition from other fibres; wool-processing mills are coming under enormous pressure to increase the quality of their products. Although most mills are acutely aware of this pressure, marketing systems in the past have not provided much inducement for mills to improve product quality. Reform of these systems to provide the necessary incentives is essential if the Chinese wool-processing industry is to remain competitive.

Chapter 3

A Woolly Story

Transforming Chinese wool marketing from part of the semi-subsistence economy of traditional nomads into a modern materials-handling system supplying valuable raw materials for a commercial economy, has been a difficult task that is far from complete. This chapter briefly outlines the evolution of wool marketing in China since 1949. The emphasis is on the major political events and institutional changes which have influenced the development of the marketing system. This sometimes turbulent past and its impact on the currently evolving elements of a market-based system are themes developed in a number of subsequent chapters. Here some of the key events and interventions in Chinese wool marketing are canvassed as a background to the later discussions.

3.1 Nomadism to Communes (1949 to 1978)

Prior to 1956, nomadic herders sold their wool during so-called "Nadam Fairs" when they gathered together once or twice each year for trading, celebrations and games. The primitive local wool markets which developed at these fairs were not subject to any form of government control. All of the wool grown in China at that time would have been very coarse by present-day standards. As outlined in Longworth and Williamson (1993, Chapter 4), it has only been since the 1950s that attention has been focused on breeding for finer wool production.

Government control over the marketing of rural products was first introduced in New China in 1951 but was not in place in most pastoral areas until 1955/56 (Fig. 3.1). Under the new unified purchase and distribution system, rural commodities were categorised into three groups, namely: category I which included the basic food grains and cotton; category II which included the important raw materials and export items from animal husbandry and horticulture including wool, cashmere, pork, mutton, beef, leather, fruit and vegetables; and category III which consisted of all other agricultural products not regarded as especially important. Under the unified purchase and distribution system and the category II classification, prices for animal products including wool were established as part of the State plan and the State allocated production quotas to be filled by each province or autonomous region at a fixed price. The quotas were set by the State Planning Commission with advice from SMCs and other organisations and varied from time to time. In general, production quotas were determined by historical average yields adjusted for seasonal factors. Although free markets for residual or over-quota production were permitted to exist, usually only a small residual was available for free trade. The unified purchase and distribution system for rural products essentially sought to achieve the level of accumulation necessary for China to modernise its urban industries.

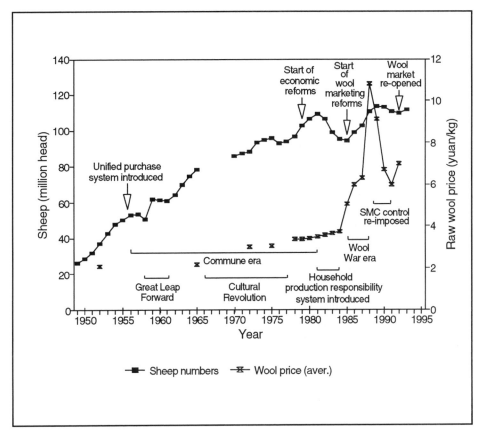

Fig. 3.1 Key Events in the History of the Chinese Wool Industry

Apart from these initial efforts aimed at controlling the market for wool, other significant events in the history of Chinese wool marketing occurred in the mid-1950s. SMCs were established under the guise of farmer cooperatives, but they soon became the vehicle by which the State effected the unified purchase and distribution system. The local government wool tax, which has subsequently proven to be a major impediment to deregulating the market for raw wool (Section 5.2.2), was also introduced at this time. On the other hand, some key elements of a commercial wool-marketing system were also emerging. The traditional "Nadam Fair" markets were simply not sophisticated enough to facilitate anything other than a direct exchange between local buyers and local sellers. As a step towards the development of a more comprehensive marketing system for wool, the FIB introduced in 1955 a set of wool samples with standard characteristics. These samples were intended to provide a rough wool-grading system.

The need for widely accepted grading standards was appreciated early in the commercial development of the Chinese wool industry. However, a considerable period of time was needed to develop and implement a set of national standards. Although wool samples with standard characteristics were widely introduced in the mid-1950s, it was not until 1976 that a national standard for the purchasing of raw wool emerged. Similarly, the industrial grading standard developed under the auspices of the Ministry of Textile Industry was not introduced until 1979, even though work on this standard had begun in the early 1960s (Section 4.4).

The introduction of the people's communes in pastoral areas and the "Great Leap Forward" between May 1958 and January 1961 led to complete closure of all free markets for pastoral products (Fig. 3.1). Furthermore, as attention focused on grain output, State prices for pastoral products stagnated. However, the failure of the "Great Leap Forward" campaigns forced the government in 1961 to develop new policies relating to the production and marketing of agricultural and animal husbandry products in an effort to encourage better production and management. Incentives for pastoral production were bolstered in many areas by the introduction of a kind of production responsibility system and free markets were reintroduced for over-quota production. The Cultural Revolution of 1966 to 1976, however, once again led to stagnation in the pastoral areas. Free markets were abolished and privately-owned animals were confiscated by the communes.

3.2 Economic Reforms (1979 to 1984)

The economic reforms that emerged from the end of 1978 have had a profound impact on the Chinese wool industry. However, it was not until 1985 that the wool market was directly affected. Marketing reforms for rural commodities and especially non-grain commodities were biased towards food production rather than fibre production. While the reforms put in place from 1979 onwards allowed the prices of surplus output of certain category I and II products to be negotiated in a free market, this did not apply to wool. The SMC retained its purchasing monopoly over raw wool, and herders received fixed State purchase prices for both quota and over-quota production. At the Third Plenary Session of the Eleventh Central Committee of the Chinese Communist Party (CCP) in 1978, these State purchase prices for pastoral products were increased. However, while the compulsory purchase prices for animal products in general rose by 22.6%, animal fibre prices rose by only 8.5%. In particular, meat prices rose substantially more than fibre prices.

Another key element of the post-1978 economic reforms was the introduction of the household production responsibility system to replace the communes (Fig. 3.1). The household production responsibility system in pastoral areas differed from that adopted in agricultural areas and varied considerably in form across different parts of the pastoral region. Nonetheless, as explained in Section 2.2.3, it was usually implemented in three stages. First, the sheep and other livestock were contracted out to individual households or to small groups of households. Second, the households were then allowed to purchase and own the livestock. Third, the pastoral lands were contracted out to the households.

Although the introduction of the household production responsibility system did not directly affect wool marketing, it had some important indirect effects. For instance, individual herders were given the opportunity to respond to the incentives implied in wool prices. Thus the prices paid for the different grades of wool, and for wool as compared with cashmere and other potential alternative products, have become major determinants of how resources are used in pastoral areas. At the same time, wool production was being fragmented. The large commune flocks were split up amongst many individual herders. Consequently, wool is now produced in much smaller and often very heterogeneous clips compared with the situation during the commune era. As already pointed out in Section 2.2.6, the emergence of an extremely large number of very small wool-growing units has led to major problems in relation to wool marketing.

Parallelling the reforms under the household production responsibility system were a series of fiscal reforms aimed at imparting more fiscal responsibility to individual economic agents (World Bank, 1988; Watson, 1989). Prior to 1980, the Central government was responsible for all fiscal revenues and therefore implicitly controlled investment and resource allocation at the local level. The fiscal decentralisation, another key plank of the post-1978 economic reforms, sought to overcome some of the more obvious inefficiencies associated with the highly centralised control of fiscal revenues. Provincial-, prefectural- and county-level governments were assigned specific tax sources and made responsible for meeting revenue targets and budgetary outcomes.

The fiscal reforms created some special problems for the pastoral region and for wool marketing. In these remote, low-income and industrially backward areas, the pressing need for local governments to develop a revenue base encouraged investment in upstream and downstream value-adding activities associated with animal husbandry. The subsequent rapid, fiscal-driven development of such activities has affected wool marketing in a number of key ways. Notably, as is outlined in Chapter 9, it changed the organisational structure of wool scouring and other early-stage wool processing in China, altered a number of the wool-marketing channels, and had a marked impact on wool quality.

3.3 Wool-Marketing Reforms (1985)

Direct changes to wool marketing were introduced in the mid-1980s. One of the consequences of the economic reforms adopted by the Third Plenary Session of the Twelfth CCP Central Committee in October 1984 was the adoption by the Central Committee and the State Council in January 1985 of the policy package known as "Ten Policy Measures for Further Reinvigorating the Rural Economy". The implementation of these new policies affected wool marketing in two ways.

First, the national unified purchase and distribution system which had existed for wool since 1956 was abolished because the Central government suspended the category II classification for wool. The policy changes in 1985 essentially replaced the "three-level" classification for rural products with a "two-level" grouping in which the purchase and distribution of some products were monopolised by the State while the marketing of all others was no longer controlled by the Central government. Wool was in the second grouping. In the case of wool, provincial governments were allowed either to set State purchase prices and control purchases and distribution or to develop a free market in the areas under their jurisdiction. Some provinces such as Liaoning allowed a free market to develop, while others such as IMAR retained tight control over wool marketing. However, even in those provinces which opted for more open market policies, such as Gansu, prefectural- and even county-level administrations sometimes overruled provincial policy and maintained closed markets. The extent to which provincial (and local) governments regulated their wool markets varied a great deal. In some cases, it involved a continuation of the unified purchase system with production quotas and set prices. In other cases, it merely involved a fee or tax on any wool exported out of the administrative area in question.

The differing levels of control over the marketing of wool in adjoining localities that emerged after the 1985 reforms were introduced created opportunities for "black markets" to develop. As a result, the marketing of wool became chaotic in some areas and created the situation described in China as the "wool war".

Second, at the same time as the Central government relinquished control over the purchasing and distribution of raw wool in 1985, it advised the governments of four major wool-growing provinces (IMAR, XUAR, Gansu and Qinghai) to develop their wool textile industries by integrating these industries more closely with the wool production sector in their province. This was the so-called "self-produce, self-process, self-sell" policy which remained in place until 1992. This policy put a great deal of pressure on these four provinces to retain some control over the marketing of wool within their borders. It was reinforced by the Central government decision to virtually eliminate the allocation of official wool import quotas to textile mills in the four provinces. These up-country mills were forced to rely on domestic wool supplies or switch to other fibres.

3.4 "Wool Wars" (1985 to 1988)

The 1985 marketing reforms led to major instability in rural areas. One outcome was the so-called "wool wars" which have been described at length by a number of authors (Watson *et al.*, 1989; Watson and Findlay, 1992). There were also "pork wars", "cotton wars", "milk wars", etc. as people tried to take advantage of the new free-marketing opportunities and as provincial and local governments endeavoured to exercise their rights to control the purchasing and distribution of these commodities. In the case of wool, the "wool wars" period was a time of volatile wool prices, wool adulteration and declining wool quality, and a generally disorderly marketing situation in many major wool-growing areas.

Like most wars, the "wool wars" were not triggered by any single event but by an unfortunate combination of a number of developments. Initially, the wars were fuelled by the rapid increase in the derived demand for raw wool which followed the increase in wool-processing capacity, especially at the local level, in the first half of the 1980s (Section 2.3). Fiscal reforms after 1980 had encouraged local authorities to develop their own revenue sources, and fibre processing was one of these sources. Local wool processing was further encouraged by the preferential treatment afforded rural-township enterprises in the "Ten Policy Measures" package of reforms introduced in 1985. Preferential terms were given in relation to both the amount which could be borrowed and interest on the loans, as well as exemption from income tax for a specified time period.

The rapid expansion in domestic processing capacity alone need not have led to marketing chaos in the domestic raw-wool sector in the late 1980s. The real trigger was the wool-marketing reforms of 1985. The Central government devolved control of the wool market to provincial governments. However, the provincial governments did not all respond in the same manner. Provinces with a comparative advantage in processing and with limited wool supplies, such as Liaoning, opened their wool markets in order to generate supplies for their processing mills. Conversely, the major wool-producing provinces had an incentive to retain a highly regulated wool market to ensure supplies to their own mills, especially given the "self-produce, self-process, self-sell" policy in force at that time. Thus competition emerged between wool-producing provinces seeking to protect supplies to their mills and the wool-processing provinces seeking to obtain wool supplies. Furthermore, the competition was not confined to provincial-level governments. Indeed, as already mentioned, in the case of provinces such as Gansu, although the province declared an open market in 1985 various prefectures/counties in Gansu (e.g. Sunan County) closed their borders in order to guarantee supplies to their recently established wool scours. Another key factor often ignored in other discussions of the causes of the "wool wars" was the desire by pastoral counties to retain control over the marketing of wool produced in the county to optimise the collection of and receipts from the wool tax. In many pastoral counties, the wool tax represents a major source of county government revenue (Section 5.2.2).

In general terms, the events of the "wool wars" period are well chronicled elsewhere. However, to emphasise the nature of the wars and to provide a flavour of what it was like "at the front line", the situation in Aohan County will be described briefly. Aohan County is in Chifeng City Prefecture in IMAR and its easternmost border adjoins Liaoning Province. (Chapter 10 in Longworth and Williamson (1993) describes Aohan County in detail.) As part of the IMAR, Aohan County operated a closed wool market policy after 1985. However, it was surrounded by three counties in Liaoning Province, namely Chaoyang, Jiangping and Beipiao, all of which encouraged free trade in wool to generate supplies for local processing or for transshipment to other mills in Liaoning.

All three Liaoning counties positioned SMC purchase stations near the border with Aohan County. Furthermore, many other government units in these counties such as veterinary stations and industry and commerce companies, as well as private dealers, actively sought to purchase wool from Aohan herders. Many Aohan herders chose to sell across the border because: the price offered was higher; the wool was purchased according to mixed grade; discounts for extra dust and sand in the wool were not as great as those imposed by the Aohan SMC; producers received 1kg of fertiliser at the State price for each 1kg of wool sold; and Liaoning buyers offered cash for the entire purchase compared with the 30% cash and 70% promissory note deal offered by the Aohan SMC. The Aohan SMC attempted to stop farmers selling their wool across the border by imposing a compulsory purchase quota on the farmers. The penalty for not meeting this quota was ¥5 for each fleece short of the quota. However, because so many farmers did not fill their quota, the officials of the Aohan SMC were reluctant to impose the penalties.

The "wool wars" were especially serious in 1986, 1987 and 1988. During these years, prices rose dramatically as did the income of herders. However, as many buyers were not experienced in wool handling and they would accept almost anything, wool quality declined. The "bubble" burst in 1989. Since the wool-buying season in China commences in June and usually only extends for a couple of months, the "June 4 Incident" in Tiananmen Square and its aftermath seriously discouraged most "illegal" wool traders from operating during the 1989 wool-buying season. In any event, the demand for wool had eased sharply during the first half of 1989 and the wool market would have been much quieter in 1989 even without the political turmoil in May and June of that year.

The "wool wars" focused attention on wool marketing and emphasised the need for change. The need for better grading and quality control became widely recognised. The Ministry of Textile Industry (representing the mills) and the Ministry of Agriculture strengthened their alliance and determination to reform wool marketing in the face of opposition from the SMCs (Ministry of Commerce). One outcome of these joint efforts was the development of wool auctions, with the first such auction being held in Urumqi in 1987.

3.5 Temporary Reinstatement of SMC Control (1989 to 1991)

As already explained, by mid-1989 the market conditions for raw wool were much less buoyant than at the beginning of the 1988 wool-buying season which had been the peak of the wool boom in China. The strong deflationary monetary and fiscal policies implemented by the Central government from September 1988 had severely dampened the demand for finished textile goods, for textile fabrics and yarns and consequently for textile raw materials like raw wool. These market developments were accentuated by the political events of May and June 1989. In fact, the "private" wool buyers and dealers who had been so active in 1987 and 1988 had virtually disappeared by mid-1989.

When the wool-buying season opened in June 1989, the market conditions in the IMAR were so bad that the provincial government decided to close the wool market in all prefectures for the foreseeable future. Once again, the grass-roots SMCs became the only organisations permitted to buy wool from farmers in the IMAR. Although the XUAR government had been able to retain more control over wool marketing after 1985 than was the case in the IMAR, the years 1987 and 1988 were particularly chaotic even in the relatively isolated XUAR. The fall-off in demand in the second half of 1989 as non-SMC buyers deserted the XUAR wool market also led to a reconfirmation of the SMC as the sole buyer in the XUAR. Thus, while the reinstatement of a monopoly buying status for the SMCs after mid-1989 was instigated by the relevant authorities as a means of reasserting government control over wool marketing, it was also related to the dearth of other buyers. In fact, without the government-financed activities of the SMCs in 1989, there would have been little opportunity for herders to sell their wool in that year.

The fall-off in final demand for wool-based goods, coupled with the unrealistically high State prices and inadequate grade-price differentials established by the Price Bureaus and imposed on the SMCs, led to a build-up of stocks of raw wool, especially in some of the lower grades of wool (Section 7.3). The emergence of large stockpiles created major problems for the SMCs both in terms of handling the wool and financing the stocks. Being unable either to sell the wool or to obtain payment for wool already sold to the higher-level SMCs and wool-processing mills placed the county and township SMCs in a difficult financial situation. For instance, five months into the 1989 wool-purchasing season in Wongniute County of Chifeng City Prefecture in IMAR, the SMC was still unable to pay farmers for their wool and was forced to issue promissory notes. Eventually, money for cashing these notes, amounting to ¥5.5 million, was provided by the Central government through the Agricultural Bank of China at an interest rate of 15% per annum. The build-up in SMC stocks in 1989 and 1990 clearly illustrated the need for more appropriate grade-price differentials (Section 5.1.2) and, in general, more flexible pricing and purchasing arrangements. The fall-off in domestic demand and the consequent build-up in stocks of domestic wool were also the key reasons behind greater control of imports from mid-1989 (Section 11.2.2).

The sudden decline in the domestic demand for wool products which followed the implementation of tough anti-inflationary fiscal and monetary policies of late 1988 affected not only the SMCs but also other units on the wool-marketing chain and elicited further policy responses. Specifically, the collapse of the domestic market for wool-based goods led to large losses in the wool-processing sector, a sector that had expanded rapidly during the 1980s. In response, MOTI under the Eighth Five-Year Plan issued an edict that there was to be no expansion in the number of spindles in the industry. Modernisation of equipment and joint-venture arrangements, however, lessened the impact of these capacity restrictions since new spindles can greatly increase spinning capacity without any increase in the number of spindles. Chapter 10 elaborates on the issues which confronted the processing sector.

Although the monopoly power of the SMCs was formally restored in the wool-producing provinces in mid-1989, some interesting experiments in relation to wool marketing in these provinces occurred after that date. In particular, mills were given greater freedom, both officially and unofficially, to deal directly with State farms which produce the more homogeneous, better-quality wool. Initially, the trials were limited to certain State farms and certain types of wool, although the mills and State farms used the opportunity to exchange other wool as well. The direct deals between mills and State farms facilitated extension of other marketing reforms, especially in relation to testing and the sale of wool on a clean-yield and industrial-grade basis. These officially and unofficially sanctioned developments seemed to be preparing the way for the eventual reintroduction of open markets.

3.6 Open Markets (from 1992)

By 1992, the demand for wool products in China had improved significantly and the stock problem had been relieved if not resolved. Consequently, at the beginning of the 1992 buying season the XUAR government took the opportunity to declare the wool market in XUAR open for the indefinite future. All prefectures and counties accepted the decision, and in theory at least, there was a free market for wool in XUAR. The IMAR followed suit, as did most provinces, prefectures and counties in the pastoral region.

In principle, subject to modest formal registration requirements, any unit or individual can now trade in raw wool in almost all parts of China. However, in practice as discussed in Section 5.2.1, not all buyers are equally welcome—the playing field is far from level! Local governments continue to favour and support their SMCs and to intervene in various, sometimes subtle, ways to ensure that the SMCs remain the dominant buyers of raw wool.

Nevertheless, the apparently permanent shift towards a free market for wool in China since 1992 has renewed the need for macro industry-wide changes in how wool is marketed. Many of these changes will create major new opportunities for agribusiness entrepreneurs.

Chapter 4

Setting Standards

Central to the commercialisation of the wool industry in China has been the development of industry standards or grading systems for raw wool and for wool tops. Chinese wool growing is now characterised by an enormous number of production units spread over a vast geographical area. Each wool-growing unit typically produces relatively tiny quantities of many different types of wool. On the other hand, wool processing occurs on a large scale with mills requiring relatively large lots of reasonably homogeneous wool. The marketing system is required to bridge the gap between growers and processors. The small quantities of wool must be purchased from growers, assembled with like lots from other growers, and eventually sold to the mills in large, relatively homogeneous parcels. An industry-wide standard which defines the various broad categories of raw wool is required to facilitate this aspect of marketing. Similarly, since wool tops are also widely traded, the marketing of tops is greatly assisted by the existence of a generally accepted set of standards for tops.

The industry standards which emerged in the 1950s, 1960s and 1970s were more-or-less satisfactory during the commune era. With a completely centrally planned economy, production quotas, administratively-determined prices, and large production units (communes), it was much less difficult to bridge the gap between wool-growing units and processing mills. Pressures to improve wool quality came from managerial directives rather than from incentives transmitted via price premiums and discounts for various grades of wool.

In the early 1980s, the communes disappeared and the household production responsibility system ushered in an era of private wool growing. Instead of a modest number of large communes and State farms producing wool, suddenly around 90% of the wool was being grown by literally millions of individual households or small groups of households. Wool marketing became even more complicated after 1985 once the Central government suspended the category II classification for wool and free markets for wool began to emerge.

Grading systems developed prior to the 1980s were no longer capable of satisfying the needs of the new marketing arrangements. However, attempts to improve the industry standards both for raw wool and for tops have built on the old widely understood but out-of-date systems rather than starting afresh. This chapter, therefore, outlines the evolution of the various standards in some detail and examines both the pressures for, and obstacles to, change.

4.1 Outline of Current Raw Wool and Top Standards

The grading of raw wool is in a state of flux. In the mid-1990s, four separate industry standards exist for raw wool. The old National Wool Grading Standard, which was previously updated in 1976, has been the traditional basis upon which raw wool was purchased from growers. It was technically replaced in December 1993 by a new National Standard. However, it will take a while for the whole industry to change to the new grading system and the old and new standards are likely to coexist (unofficially) for some time. In any event, the new National Standard incorporated only modest changes, some of which apply to types of wool not commonly grown in many parts of China. The next section outlines the old 1976 standard to indicate the essence of the grading system by which raw wool has been traditionally purchased in China. The changes embodied in the new standard implemented in 1993 are then highlighted in Section 4.3.

Neither the old nor the new National (purchase) Standard creates grades which are entirely appropriate for the mills, especially when the wool is to be made into tops. Consequently, the former Ministry of Textile Industry (MOTI) encouraged the development of a separate Industrial Wool-Sorts Standard. This grading system is outlined in Section 4.4. The degree to which the raw-wool grades established under the old National (purchase) Standard correspond with the industrial grades defined in the Industrial Standard, is examined in Section 4.5. No data are yet available on the correspondence between the new National (purchase) Standard and the Industrial Standard but there is little prospect of any significant improvement.

The fourth set of standards currently in use for raw wool are embodied in the Quality Standard for Auctioned Wool. With the introduction of wool auctions in 1987, the National (purchase) Standard in force at that time was considered an inadequate grading system for the better-quality wool which was expected to be sold through the auctions. The standard which is applied to wool destined for the auction marketing channel is discussed separately in the chapter on wool auctions (Chapter 8).

As with the old National (purchase) Standard for raw wool, the traditional Wool Top Standard was recently modified. Both the old and the new Top Standards are outlined in Section 4.6 and some of the difficulties encountered by those who wished to reform the old standard are discussed.

4.2 Old National Wool Grading Standard

A translation of the old National Wool Grading Standard which has been used as the basis for raw-wool purchasing since 1976 is presented in Appendix A. Given the heterogeneity of Chinese wool, this set of standards is a very simple grading system. Wool is basically classified into 10 categories. A distinction is first made between fine wool and semi-fine wool based on fibre fineness as measured by quality number or spinning count (Fig. 4.1). As explained in more detail in Appendix A, the old standard specifies fine wool as having a count of 60s or more and semi-fine wool as having a count of between 46s and 58s. Each of these two groups is then further subdivided according to whether the wool is homogeneous or otherwise. In the latter case, it is referred to as improved wool. Each of the four wool types (fine wool, improved fine wool, semi-fine wool, improved semi-fine wool) is then split into two quality grades, namely Grade I and Grade II. Although the standard lists a range of factors that separate the two grades, the key differentiating factor is staple length. The division between Grade I and Grade II for fine wool is 6cm (with 6cm and above being Grade I) and for semi-fine wool it is 7cm. While not formally specified in the old standard as separate grades, homogeneous wool exceeding a particular length is classified as special grade wool. In the case of special grade fine wool, the staple length must be 8cm or more, while special grade semi-fine wool requires a staple length of 10cm or greater.

The basic eight grades shown in Fig. 4.1 and the two special grades of homogeneous wool complete the 10 categories by which fine and semi-fine wool is purchased from the growers. Of course, a great deal of wool with a count lower than 46s is produced in China. The old National Standard does not cover this wool which is generally described as coarse/local wool.

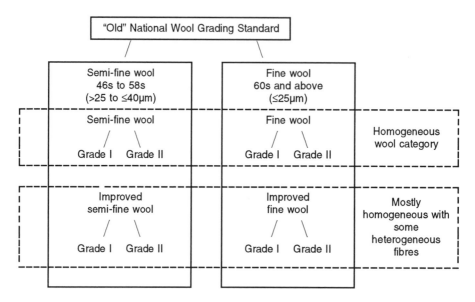

Fig. 4.1 Categories of Wool Established by the "Old" National Wool Grading Standard

An understanding of the development of the old National Standard will help explain its apparent elementary nature. The origins of the old standard extend back to 1950 when the Fibre Inspection Bureau (FIB) was first set up within MOTI as a quality control unit for Chinese textiles. One of the first commercial regulations established by the FIB at this time was the definition of a fine wool sheep. A sheep was defined as being of fine wool type if its fleece was 60% or more fine wool (wool of a 60s count or better) and of coarse wool type if it had less than 60% fine wool. This commercial breed standard established by the FIB in 1950 allowed for only two types of wool—fine wool and coarse wool. Prior to 1950 there were no clearly defined categories of sheep according to wool type.

In 1955, in response to criticism from growers and processors over the subjective nature of wool grades at the time of purchase and sale, the FIB introduced wool samples with standard characteristics. In 1957, the FIB extended the fine wool and coarse wool standards to include improved wool standards. Under these standards, improved wool was defined as being wool produced from an F_1, F_2, F_3 or F_4 crossbred sheep produced by using pure fine-wool rams over native or local-breed ewes. Grades of improved wool were inversely correlated to each of these four crosses (i.e. Grade IV improved wool was defined as being produced from an F_1 crossbred sheep; Grade

III improved wool was defined as being produced from an F_2 crossbred sheep, and so on up to an F_4 crossbred sheep which produced Grade I improved wool).

While the 1957 standards were an improvement on the earlier wool samples and were a major step forward compared to the situation when there were no standards whatsoever, they were not sufficiently well developed to meet the rapid changes in wool production and wool processing which occurred over the following two decades. In particular, they were not well suited to accommodating or encouraging the improvement in wool quality brought about by enhanced sheep-breeding activity, and the expansion in wool processing. Chinese policy-makers and wool industry officials appreciated the need to review the standards in order to serve better the needs of both the wool-producing and wool-processing sectors. In 1975, a revision of the standards was commenced in accordance with the instructions of the State Council and State Planning Commission.

The principles upon which the new standard was to be based, and which are set out in full in Appendix A, illustrated the "ambitious but achievable" approach. The new standard was intended not only to facilitate raw-wool purchasing in the domestic market but also to achieve the more idealistic goals of increasing wool production and quality as well as promoting the development of wool processing. However, there were several serious pragmatic constraints in place. In particular, while the need to develop quantifiable indicators of wool quality in line with international standards was recognised, the technical sophistication of the majority of herders was undoubtedly much lower than required for the adoption of international standards. Furthermore, since the new standard was expected to be developed by consensus and to have universal support so that it would be adhered to on a nationwide basis, there was an overriding pressure to adopt a lowest common denominator approach.

Although the intent may have been to meet the needs of both wool producers and wool processors, the final outcome of the review was more closely aligned with the interests of the former. The State Standards Bureau was given responsibility for the development of the new standard and established a wool standard investigation team involving representatives of the Ministry of Agriculture (MOA), MOTI and the All China Union of Supply and Marketing Cooperatives. The team conducted most of their investigations between October 1975 and April 1976 and concentrated on the wool-producing pastoral region. Rather than being a total rewrite of the pre-existing standard, the new standard developed by the investigation team was heavily based on the old 1957 standard.

Although the new standard was designed to facilitate sheep-breeding efforts and the grading of raw-wool supplies to mills, it fell well short of both these goals, especially the latter. It seems the standard was meant to assist processors by identifying homogeneous fine wool suitable for combing and worsted manufacturing, semi-fine wool which would be appropriate for yarn production, and improved (not sufficiently homogeneous) wools for carding. However, this three-way classification was obviously too broad and of only limited value to the processors. Consequently, a completely different standard was developed by MOTI at about the same time to service the needs of mills. This mill-specific standard, known as the Industrial Wool-Sorts Standard, is described in Section 4.4.

The National Wool Grading Standard introduced in 1976 became the framework on which State prices were listed (see Chapter 5). The standard also specified other conditions which could influence the net price received for any particular lot of wool. For example, the standard specified that wool with an average staple length of less than 4cm (where shearing was conducted twice a year) should incur substantial price penalties (50%). Head, leg and belly wool was to receive only 40% of the price of corresponding fleece wool, while yellow-stained wool and wool branded with tar or paint were to receive discounts of between 35 and 40%. The National Standard also listed inspection and sampling procedures as well as packing and labelling requirements (see Appendix A).

As will be outlined in subsequent chapters, the National Wool Grading Standard introduced in 1976 created a dilemma for the Chinese wool industry. On the one hand, the standard was simply too elementary to be of much value to the mills which were often forced to bear the additional costs of re-sorting. Moreover, it had insufficient grades to allow appropriate premiums and discounts for quality. However, on the other hand, even with this simple standard it has proven extremely difficult to avoid incorrect grading and other abuses of the grading system.

4.3 New Raw-Wool Purchasing Standard

The new raw-wool purchasing standard was issued by the National Supervisory Bureau of Technology in April 1993 and was supposed to be implemented throughout China from December 1993. Full details of the new National Wool Grading Standard may be found in Appendix B. Table 4.1 presents a comparison of the technical requirements for the various grades as formally stipulated in the old and the new purchasing standards. The new standard was intended to overcome some of the limitations of the 1976 standard. While there are some important technical differences between the old and the new standards, the new standard is not likely to improve significantly the usefulness, from a processing perspective, of grading at the time of purchase.

Probably the most substantial change is that the new standard explicitly decouples the definition of improved fine wool from fine wool (and improved semi-fine from semi-fine wool). Indeed, the new standard does not make a distinction between improved fine and improved semi-fine wool—there is just one category "improved wool". In this respect the new purchasing standard is in line with the Auction Standard (see Section 8.2.1). The practical significance of this change will be to encourage the market to distinguish more carefully between what is called "homogeneous wool" and improved (but not completely homogeneous) wool. Under free market conditions, the price premium for genuinely homogeneous wool is likely to be larger than has traditionally been the case in China.

The drawing of a sharper distinction between homogeneous wool and improved wool could have major ramifications since the vast majority of private households raising sheep produce improved (i.e. crossbred) wool. Indeed, it may prove to be a major obstacle to having the new standard accepted at the grass roots. It also suggests that there is an official intention to cease the longstanding and misleading practice of presenting Chinese wool production statistics grouped into just two categories (fine and improved fine/semi-fine and improved semi-fine). These categories are usually simply labelled "fine" and "semi-fine" despite the high proportion of improved wool in both categories. If this is the case, it will allow Chinese policy-makers and others to monitor more easily the upgrading of the bulk of the Chinese clip from improved (or crossbred) wool to genuine fine and semi-fine wool. (See Longworth and Williamson, 1993, pp.51–52 for more detail on this important point.)

Another significant technical change incorporated in the new standard is that mean fibre fineness is explicitly expressed in microns as well as spinning count. Perhaps more importantly, the definition of semi-fine wool has been extended to include wools with counts from 58s down to 36s (Section 4.1 in Appendix B). Special grades for fine and semi-fine wool are now explicitly included. Furthermore, the special grades have been subdivided into a number of fineness classes, some with different staple-length requirements (Table 4.1). The additional details for special grade wool are probably designed to assist in bridging the gap between the purchasing standard and the Industrial and Auction Standards in relation to technical details for top-quality wool. As explained in Section III.1(1) of Appendix A, special grade was not formally included in the old standard presumably because so little of the Chinese clip at the time the old standard was developed in the mid-1970s was expected to be wool of this type. Nowadays, while the proportion of the total clip which would be classed as special grade is still low, some particular areas produce substantial quantities of special grade wool.

The more detailed classification of fine and semi-fine wool of Grade I according to fineness (and, in the case of semi-fine wool Grade I, according to length as well) will be of significance to a much larger group of producers. However, the new standard (as with the old standard) still defines the grades essentially on the basis of staple length, with fineness and the other technical specifications being "reference indicators". (See Section 4 in Appendix B.)

Table 4.1 indicates that there have been some other technical changes besides those relating to fineness and length. For example, one relatively minor difference between the old and the new standards concerns the grease and suint requirements. In the old standard, this reference indicator was defined in terms of the number of centimetres of wool from the bottom of the staple which contained grease and suint. Section II.3(4) of the old standard (Appendix A), however, noted that the 3cm length was a minimum requirement and that to produce the best-quality products and to achieve greatest processing efficiency the grease and suint had to cover more than two-thirds of the length of the staple. The new standard sought to move away from the minimalist approach to one encouraging product improvement by stating the grease and suint requirement in terms of the proportion of the staple length which should contain grease and suint.

Perhaps one of the most significant features of the new standard is that it contains no reference to "quality differentials". These fixed ratios have dominated the setting of administered prices by the State in the past (Section 5.1.2). Even under the pseudo-free market conditions between 1985 and 1988, they clearly had a big influence on the premiums and discounts paid for the various grades. Furthermore, as will be discussed in Chapter 8, the starting and reserve prices set for each lot in the fledgling auction system have been determined according to quality differentials in much the same way as State purchasing prices have been traditionally determined. Deleting quality differentials from the new standard, therefore, is a major break with the past.

At the grass roots, implementing the new standard will require only modest adjustments. The new licensing system for wool buyers requires that individuals must attend training programs and obtain the necessary certificate before being permitted to buy wool for the SMCs, for any other government unit, or for private traders. These arrangements should ensure that within a reasonable timeframe wool buyers will have incorporated the changes in the new standard into their operations. Unfortunately, however, the new purchase standard does little to address the major problems with

Table 4.1 A Comparison of the Technical Requirements in the "Old" and the "New" National Wool Grading Standards

Wool type	Grade	Fineness (μm and count)	Natural length (mm)	Grease/ suint*	Morphological characteristics	Quality differential ratio (%)
Fine						
	(a) Old Standard (implemented in 1976)					
	I	60s and above	60–79	≥3cm	All the wool should be naturally white and homogeneous fine wool, uniform in fineness and length of wool staple, normal crimpness, soft-feel, with elasticity even at the tip of the wool. Part of the tip of the staple can be dry or thin wool, but without withered hair and kemp.	114
	II	60s and above	40–59	<3cm	Has the same quality features as Grade I, but the length and grease and suint are less than Grade I; or the length and grease are the same as Grade I but the fineness and uniformity are less than Grade I, with loose and open staple, subnormal crimpness, and poor elasticity.	107
	(b) New Standard (implemented December 1993)					
	Special	18.1–20.0 (70s) 20.1–21.5 (66s) 21.6–23.0 (64s) 23.1–25.0 (60s)	≥75 ≥75 ≥80 ≥80	≥50** ≥50 ≥50 ≥50	No presence of coarse/withered/kemp wool. All homogeneous and naturally white wool with similar fineness and length and normal crimps. Possible presence of small sharp wool tips on some staples.	N/A
	I	18.1–21.5 (66–70s) 21.6–25.0 (60–64s)	≥60 ≥60	≥50 ≥50	The same as for special grade but with the possible presence of withered wool at the top of some staples.	N/A
	II	≤25.0 (≥60s)	≥40	+***	No presence of coarse/withered/kemp wool. All homogeneous and naturally white wool with slight difference in fineness of wool fibres. Loose structure of staples.	N/A
Semi-fine						
	(a) Old Standard (implemented in 1976)					
	I	46s to 58s	70–99	With grease and suint	All the wool should be naturally white and homogeneous semi-fine wool with uniform fineness and wool length. Light and large crimpness, good elasticity, nice lustre, plain wool tip or small sharp wool tip and small hair plait with the shape of plied yarn. Thick hair plait for coarse semi-fine wool, but without withered hair or kemp.	114
	II	Ditto	40–69	Ditto	Ditto	107

Table 4.1 A Comparison of the Technical Requirements in the "Old" and the "New" National Wool Grading Standards (*continued*)

Wool type	Grade	Fineness	Natural length	Grease/ suint*	Morphological characteristics	Quality differential ratio
		(μm and count)	(mm)			(%)
	(b) *New Standard (implemented December 1993)*					
	Special	25.1–29.0 (56–58s)	≥90	+	No presence of coarse/withered/kemp wool. All homogeneous and naturally white wool with similar fineness and length, and big and shallow crimps. Glossy staples with plain or small sharp tips or with some small wool plaits at the wool tips.	N/A
		29.1–37.0 (46–50s)	≥100	+		
		37.1–55.0 (36–44s)	≥120	+		
	I	25.1–29.0 (56–58s)	≥80	+	Ditto	N/A
		29.1–37.0 (46–50s)	≥90	+		
		37.1–55.0 (36–44s)	≥100	+		
	II	≤55.0 (≥36s)	≥60	+	No presence of coarse/withered/kemp wool. All homogeneous and naturally white wool.	N/A
Improved	(a) *Old Standard (implemented in 1976)*					
	I		Literal description only (See Sections III.1 and III.2 in Appendix A)			100
	II		Literal description only (See Sections III.1 and III.2 in Appendix A)			91
	(b) *New Standard (implemented December 1993)*					
	I	Fineness not specified	≥60	+	Presence of ≤1.5% coarse/withered/kemp wool. All naturally white and basically homogeneous improved wool. Staples consist of fine hair and heterotypical wool. The homogeneity, crimps, grease/suint content and morphological characters are inferior to fine and semi-fine wool. Presence of small or medium-sized wool plaits in staple.	N/A
	II	Fineness not specified	≥40	+	Presence of ≤5.0% coarse/withered/kemp wool. All naturally white and heterogeneous improved wool. Staples consist of more than two types of wool fibre with big or very shallow crimps. Presence of small or medium-sized wool plaits and grease/suint in staples.	N/A

N/A, Not applicable.
* In the old standard, the length of the staple up from the bottom which must contain grease and suint was specified in cm. In the new standard, the requirement is for a certain proportion of the staple to contain grease and suint.
** Percentage of the total length of staple wool containing grease and suint.
***The staple must contain some grease and suint.
Source: Derived from Appendices A and B.

traditional raw-wool purchasing practices in China. In particular, it does not adequately tackle the issues of either fibre tensile strength or trading on the basis of clean yield.

Certainly, the new standard states that special grade and Grade I fine wool and semi-fine wool should not contain tender wool (Section 4.12 in Appendix B). The new standard also requires that if the weight of the batch (or lot) of wool being traded exceeds 2,000kg (greasy), then the wool must be officially tested and the sale price determined on a clean-wool basis (Section 6.4, Appendix B). But neither the stricture about tender wool nor the requirement that large lots be objectively measured is likely to affect the way purchases are made at the grass roots. The new standard does not spell out precisely what constitutes tender wool. In addition, virtually no private households would be offering for sale lots as large as 2,000kg or more.

4.4 Industrial Wool-Sorts Standard

At the same time as the old National Wool Grading Standard was being developed in 1975/76, a parallel standard was being created under the auspices of MOTI. The second standard was necessary because grading under the National Standard did not provide sufficiently homogeneous wool to make good-quality tops. The MOTI standard, known as the Industrial Wool-Sorts Standard, was designed so that mills could make best use of the "unsorted" wool purchased according to the old National Wool Grading Standard.

An industrial wool-sorts standard first appeared in 1960 but was of limited value to the mills. As with the National Wool Standard which existed at the time, it was recognised that further development was necessary but, as explained in Appendix C, progress was slow. In 1969, MOTI called upon the textile bureaus in the major east coast processing centres of Shanghai, Beijing and Tianjin to set about revising the old standard. The revised standard was then tested at some of the major mills. However, it was not until 1975 that a thorough review of the revised standard was conducted. This review was slightly broader in scope because it included textile authorities from the wool-producing regions as well as other key processing regions such as Liaoning. The Industrial Wool-Sorts Standard was finally issued in 1979.

The Industrial Wool-Sorts Standard is listed in Appendix C. Under this standard, the wool is first split into two categories, count wool and sort wool, according to the homogeneity of the wool (see Appendix C for full details). Count wool is basically homogeneous wool which is further classified into count ranges depending on the fineness of the wool. Within the count wool, there is a series of specifications for each category of count wool, including average fineness, fineness discreteness, rate of coarse-cavity wool, proportion of staple length which contains grease and suint, and staple length. However, fineness discreteness, rate of coarse-cavity wool and staple length with grease and suint are all only reference indicators. That is, they serve only as a reference for mills and not as strict sorting criteria. Apart from average fineness, staple length is the only critical classifying criterion and there are four staple-length categories. The proportion of wool shorter than the lower limit of the staple length of any particular class should not exceed 15%, of which 5% must not be shorter than the lower staple-length limit of the next lowest class.

Sort wool is the more heterogeneous wool and is further divided into six sort classes depending on the percentage of coarse-cavity wool. Although average fineness is also specified for the different sort classes, this again serves only as a reference indicator.

Apart from count and sort wool, the specifications require that all defective wool as well as coloured wool be removed from the fleece and be treated and grouped separately.

4.5 Compatibility of the Old National Wool Grading Standard with the Industrial Wool-Sorts Standard

The lack of correspondence between purchase grades (according to the old National Standard) and the grades of wool required for processing usually meant that mills had to re-sort the wool, especially if the wool was to be made into tops. Tables 4.2 to 4.4 present data which illustrate the extent of the divergence between the wool purchase grades and the industrial grades. Table 4.2 shows the distribution of all raw wool which was regraded from purchase grades into industrial grades at the IMAR topmaking mill in Huhehot in 1991. Comparisons of that sorting are then made with the same mill in a different year, namely 1989 (Table 4.3), and with another mill, Gansu No. 1 mill, in the same year (Table 4.4). In Table 4.5, the distribution into industrial grades of sequentially different batches of raw wool for Xinjiang No. 1 mill over the period July 1991 to May 1992 is presented.

Tables 4.2 to 4.4 highlight the wide dispersion over industrial grades of any particular purchase grade. Three-fifths of the wool regraded by the IMAR topmaking mill in 1991 was fine wool Grade II. Although one-half of this wool fell into the 66s count industrial grade, one-eighth of it was also Sort I wool while a similar amount was woollen-grade wool. Indeed, a much lower though still important share (3.7%) was actually mixed-sort wool with vegetable fault. Fine wool Grade I (10.5% of all raw wool regraded), although containing a higher proportion of 66s and 64s count wool (74% in total), still had significant quantities of Sort I wool, woollen-grade wool and mixed-sort with vegetable fault. The other major grade of raw wool regraded, improved fine wool Grade I (accounting for 14.1% of raw wool regraded), exhibited even greater variation. Indeed, the major industrial type that this wool sorted into accounted for only one-quarter of the wool sorted, with the remaining wool sorting into many different industrial grades. It would be almost impossible for the mill to know *a priori* how this improved fine wool Grade I wool would have sorted out among the different industrial grades. Even the more "homogeneous" categories of purchase grades such as special fine wool had significant quantities of Sort I and mixed-sort wool in addition to 66s and 70s count wool. The semi-fine purchase grades mirrored to some extent what happened with the fine wool grades. Half of the semi-fine wool fell into 60/58s but there was almost another third re-sorted as 64/66s count wool. Since the definition of semi-fine in the old National Standard is 46/58s count wool (Appendix A), the data in Table 4.2 provide evidence that, at least in 1991, the purchase standard was being applied conservatively by the SMCs.

Table 4.3 presents the results of re-sorting raw wool into industrial grades at the same mill two years earlier. Once again, each of the raw grades distributed into many different industrial grades. A relevant issue, however, is whether the distribution remains roughly the same across the years. In general, there is a broad similarity in the distribution patterns shown in Tables 4.2 and 4.3, while some grades such as special grade fine wool exhibit an especially close correspondence. However, for the main purchase grades there are important differences between years. For instance,

Table 4.2 Re-Sorting of Wool from Purchase Grades to Industrial Grades: IMAR No.1 Topmaking Mill, 1991

Industrial grade	Raw wool purchase grade						
	Special grade fine wool	Fine wool		Improved fine wool		Semi-fine	Improved semi-fine Grade I
		Grade I	Grade II	Grade I	Grade II		
	------ (percentage by weight on a clean scoured basis) ------						
Count wool							
70s	6.60	0.66				14.92	9.53
66s	72.68	61.56	50.62			16.40	2.16
64s	3.96	12.38	2.75	13.13		50.82	0.60
60/58s						0.82	
56s						0.05	
40/48s							
Sort wool							
Sort I	5.50	4.38	16.17	26.60	3.59	2.61	27.28
Sort II	0.35	0.89	2.95	14.22	7.94	0.49	12.33
Sort III		0.45	1.13	6.47	7.61		1.30
Sort IV	0.08	0.27	1.20	6.64	16.20	2.20	11.65
Sort IV(a)	0.01	0.25	0.51	4.82	33.68	0.88	8.42
Other wool							
Woollen-grade	0.42	7.93	16.41	15.72	12.51	3.14	16.43
Mixed-sort, veg. fault	5.73	4.83	3.69	3.08	3.79	2.14	2.78
Mixed-sort not white	0.15	0.36	0.50	2.66	8.74	1.10	2.10
Defective wool	0.20	0.58	0.48	0.57	1.35	0.23	0.49
Yellow-stained wool	0.54	0.54	0.01	0.02			
Scabby wool	0.11	0.16	0.07	0.08		0.08	0.04
Cotted wool	0.75	0.93	0.28	1.65	0.94	0.22	0.89
Dirty-cotted wool	1.19	0.15	0.25	0.11		0.39	0.39
Dust	2.27	3.68	2.97	4.22	3.64	3.51	3.62
Total	100 (1.0)†	100 (10.5)	100 (62.3)	100 (14.1)	100 (0.5)	100 (7.7)	100 (3.9)

†Figures in brackets are the percentage of total wool sorted which fell into this grade.
Source: Topmaking Mill, IMAR (personal communication, 6/6/1992).

Table 4.3 Re-Sorting of Wool from Purchase Grades to Industrial Grades: IMAR No.1 Topmaking Mill, 1989

Industrial grade	Raw wool purchase grade								
	Special grade fine wool	Fine wool		Improved fine wool		Semi-fine wool		Improved semi-fine wool	
		Grade I	Grade II	Grade I	Grade II	Grade I	Grade II	Grade I	Grade II
	-----------	-----------	-----------	(percentages by weight on a clean scoured basis)	-----------	-----------	-----------	-----------	-----------
Count wool									
70s	8.86	0.22							
66s	67.21	48.22	52.83						
64s	4.35	33.64	12.91	15.74	0.36	28.66	51.17	3.64	0.34
B 64s		1.91	5.31						
58s/56s						46.65	10.72		
40s/48s						2.31	0.23		
Sort wool									
Sort I	5.24	2.42	11.51	37.42	7.06	7.39	17.60	33.49	4.59
Sort II	0.54	0.21	1.09	12.95	10.20	1.80	2.96	22.69	10.78
Sort III	0.18	0.05	0.43	5.77	9.55	0.59	0.90	10.07	14.60
Sort IV(A)	0.35	0.04	0.38	4.13	18.63	0.57	1.19	7.21	25.29
Sort IV(B)				0.96			0.11	0.42	11.53
Sort V				1.34	32.26	0.48	0.56	4.24	14.26
Other wool									
Mixed-sort not white	0.83	0.03	1.05	3.68	6.62	1.63	3.31	4.01	5.96
Mixed-sort, veg. fault	4.55	4.72	4.49	6.12	1.32	4.10	4.34	6.11	7.04
Short wool	4.49	1.45	5.80		1.60	3.96	3.96		
Edge wool		1.08	5.80	8.96	4.64	0.05	0.05	6.03	1.32
Cotted		0.03	0.45	0.33	4.80	0.16	0.18	0.08	1.07
Mite-affected		0.01	0.05	0.03	0.08	0.19	0.01	0.02	
Yellow-stained wool		0.36	0.03		0.02		0.01	0.11	0.08
Crutching		0.19		0.14			0.17		1.29
Dust	3.40	4.82	3.67	2.42	2.26	1.46	2.54	1.88	1.85
Total	100.00	100.00	100.00	100.00	100.00	100.00	100.00	100.00	100.00

Source: Topmaking Mill, IMAR (personal communication, 29/5/1990).

Table 4.4   Re-Sorting of Wool from Purchase Grades to Industrial Grades:
Gansu No.1 Mill, 1991

Industrial grade	Raw wool purchase grade			
	Fine wool		Improved fine wool	
	Grade I	Grade II	Grade I	Grade II
	- - (percentage by weight on a clean scoured basis) - -			
Count wool				
64s	46.90	35.40	16.17	11.40
60s	6.13	16.20	1.32	1.22
Sort wool				
Sort I	23.40	20.10	17.89	15.40
Sort II	6.40	11.40	16.53	35.40
Sort III	3.01	4.08	25.26	4.50
Sort IV			14.11	16.40
Other wool				
Light black-and-white fine wool	12.10	10.39	1.05	10.27
Dark black-and-white fine wool	0.60	0.35	3.15	2.40
Cotted wool	0.35	0.70		
Dirty cotted wool	0.40	0.50	2.11	1.40
Wool with vegetable debris	0.71	0.88	1.62	0.41
Light black-and-white coarse wool			0.79	1.20
Total	100.00	100.00	100.00	100.00

Source: Gansu No. 1 Mill (personal communication, 8/6/1992).

only one-eighth of the fine wool Grade I in 1991 sorted into the 64s count industrial Grade whereas in 1989 the corresponding figure was one-third. Similar differences were exhibited by the data for improved fine wool Grade I. On the basis of Tables 4.2 and 4.3, it seems that the mill managers concerned could not confidently predict the industrial-grade composition of any particular purchase grade even when this composition is averaged out over many different lots purchased during a year. Of course, the expected variation in any two lots of wool of the same purchase grade would be even greater.

Table 4.4 provides a cross-sectional comparison with Table 4.2 by examining the wool sorting at the Gansu No. 1 mill in 1991. Once again, the table reveals that the raw-wool purchase grades fall into a wide range of industrial grades. For instance, no single industrial grade accounted for at least half of the fine wool Grade I purchased by the Gansu mill (Table 4.4). Almost one-quarter of the fine Grade I wool re-sorted into Sort I while one-eighth was light black-and-white fine wool. Indeed, fine wool Grade II and improved fine wool Grades I and II each sort into no less than five industrial-wool grades accounting for more than 10% of the raw-wool grade. More notably, the distribution of the fine wool and improved fine wool grades into the industrial grades bears no resemblance to that at the IMAR topmaking mill. Some of the variation undoubtedly reflects the different sources of the raw wool. But it is also an indictment on the broad nature of the raw-wool grades and the general lack of correspondence between the raw-wool grades and the industrial-wool grades.

Table 4.5 Wool Sorting to Industrial Grades for Xinjiang No. 1 Mill, July 1991 to May 1992

Industrial grade	Batches of wool sorted (as defined by dates)							All wool sorted
	8/7/91–31/7/91	31/7/91–3/9/91	10/10/91–27/11/91	29/11/91–25/1/92	25/1/92–24/3/92	26/3/92–25/4/92	27/4/92–27/5/92	
	- - - - - - - - - - - - - - - - - - (percentage of raw wool) - - - - - - - - - - - - - - - - - - -							
Count wool								
66s	48.28	46.31	28.88	28.49	30.81	38.18	39.44	35.44
64s	31.11	40.53	54.21	53.19	48.22	45.96	45.7	47.21
60s		0.21						0.03
Sort wool								
Sort I	3.39	2.11	3.65	1.22	2.86	1.89	2.03	2.42
Sort II–III	0.18	0.13	0.43	0.27	1.28	0.15	0.18	0.25
Other wool								
Dirty cotted wool	0.03	0.17	0.14	0.17	0.02		0.77	0.17
Vegetable matter	10.28	5.66	6.34	6.01	7.56	6.9	4.74	6.61
Black-and-white wool	0.73	0.17	0.26	0.21	0.19	0.21	0.16	0.21
Yellow-stained wool	0.07	0.11	0.23	0.07	0.15	0.26	0.13	0.15
Cotted wool	0.41	0.38	1.17	0.22	1.33	0.77	1.07	0.81
Short wool	1.68	0.71	0.76	0.71	0.85	0.26	0.54	0.74
Pitch (tar) wool	2.07	2.1	2.3	2.56	3.04	2.97	3.73	2.74
Coloured wool	0.21							0.01
Dust	1.56	1.41	1.63	6.88	3.69	2.45	1.51	3.21
Total	100.00	100.00	100.00	100.00	100.00	100.00	100.00	100.00

Source: Xinjiang No. 1 Mill (personal communication, 13/6/1992).

Table 4.5 does not directly indicate correspondence between raw-wool purchase grades and industrial grades. Rather it shows how different batches of wool (combined over all wool grades) for the Xinjiang No. 1 mill were distributed among industrial grades. The data in Table 4.5 show that the raw-wool batches were distributed over a wide band of industrial grades and that the distribution varied from batch to batch. Although the batches undoubtedly consisted of different proportions of raw-wool grades, the data suggest that mills in the XUAR are likely to experience similar problems to their counterparts in IMAR and Gansu.

The lack of any simple pattern of correspondence between raw wool purchase grades and industrial grades of wool and the variation in these patterns both over time and over space have clearly been a major problem. Not only do they emphasise the need for regrading prior to first-stage processing (an issue stressed in Chapter 9), but also they mean that prices (values) placed on the wool when purchased are unlikely to reflect accurately the true value of the wool to the mills (see Chapter 5).

Furthermore, for the reasons discussed in Section 4.3, the new wool purchase standard which came into effect from December 1993 is not likely to make any dramatic difference, especially in the short run. Several times already it has been suggested that the need to re-sort wool at the mill imposes a cost on the processor, but this is not the most serious dimension of the problem. There are two other less obvious but more important consequences. The first concerns the degree to which the marketing system can transmit the correct price signals to wool growers. At present, there is a desperate need for herders to receive incentives to produce better-quality wool. However, since the purchase grades are so broad, even a properly functioning free market will only generate a kind of weighted average price for the expected range of industrial grades embodied in the purchase grade. Leaving aside the obvious high degree of uncertainty about the actual industrial-grade composition of any lot of wool which has been graded according to the old or even the new purchase standard, the mixed average price paid to growers will have little information content. The real premiums and discounts the mills are prepared to pay for the various industrial ("true") grades will not be revealed to producers.

The second serious consequence is even more insidious. As will be discussed in Chapter 9, a growing proportion of the domestic clip is being processed at county scours before being sold to mills as clean scoured wool. Since it is virtually impossible to re-sort wool after it is scoured, it means that these local scours must accurately re-sort the wool before scouring it if it is to retain its full value to mills. Since graders at local scours are unlikely to be as highly skilled as their counterparts at the large integrated mills, local scouring may seriously downgrade the value of better-quality wool. As a result, there will be less incentive for herders to improve the quality of their wool and the future improvement of the domestic wool-growing industry will be further retarded.

4.6 Wool Top Standard

The Wool Top Standard is a mill-based standard issued by MOTI. Up until mid-1993, the standard was in line with the Industrial Wool-Sorts Standard with a classification based on the count or sort of the wool from which the top was manufactured. However, the recently revised standard has introduced some important changes. The old top standard is shown in Appendix D while the new top standard is reported in Appendix E.

Under the new Wool Top Standard, the tops are first classified according to their fineness (micron ranges). They are then graded into two types (Grade I and Grade II) according to the physical properties of the top and morphological faults. The physical properties considered are weighted average length of fibres, percentage of short fibres below 30mm and percentage of weight non-uniformity. In addition, fineness variation and length variation of the wool are used as reference indicators. Morphological faults include wool grain, wool piece and grass debris. The grade of the top is set in accordance with the lowest grade of the batch and any top that does not meet the Grade II specifications is classified as out-grade. Although, in principle, out-grade top is not allowed to leave the mill, special agreements (contracts) between the topmaker and subsequent purchaser in regard to ungraded top are sanctioned.

The events surrounding the implementation of the new Wool Top Standard present some insight into the problems and ways of introducing marketing reforms in China. Compared with the National Wool Grading Standard, it could be expected that the top standard would face fewer development problems, given the smaller number of users with more homogeneous interests than in the case of the wool purchase standard. Nevertheless, the introduction of the new Wool Top Standard has followed a tortuous path, while the drafting of the standard also reflects the difficulties even a relatively straight forward reform will encounter.

MOTI requested a revision of the Wool Top Standard as early as 1987. A consultative meeting for the revision of the National Top Standard of Domestic and Imported Wool was organised by the Shanghai Research Institute for Wool and Jute Textile Sciences and Technology and was held in Xitang, Wuxi, in October 1989. (See Appendix E, First and Second Supplements.)

One of the key changes mooted was the use of fineness, rather than spinning count, as an indicator for top classing to bring the standard in line with practices in other countries and to introduce a more objective form of measurement. Specification of the number of wool grains per unit weight was introduced to meet the needs of worsted fabric and knitting-wool fabric, and the amount of grass debris was reduced in an attempt to improve product quality. In addition, other pragmatic considerations were raised. For instance, while short wool below 30mm reduces the quality of knitting wool, because of the shortage of wool at the time the proposals were being drafted in 1988/89 the rate of short wool permitted was increased from 3 to 3.5% for 26 to 31µm wool.

Various characteristics were also proposed as auxiliary indicators, such as wool net cleanness, on the basis of difficulties in measurement or in establishing a standard specimen. Others, such as less than 1% lanolin in the top were listed as reference indicators for top quality because of their importance in spinning and further processing of tops. A discrepancy existed in using length and fineness discretion as indicators of top quality in various regions. However, as they are not used in other countries and their relationship with spinning remains to be established, it was decided that they also be used as reference indicators only.

The process of implementation was that the National Top Standard Revision Committee was to modify the standard based on the opinions of delegates and based on trial performance of the original draft. After the modifications, the new standard was to be submitted to MOTI for its examination and approval. The new Wool Top Standard was to be implemented on a whole-country basis from 1 July 1993.

4.7 Concluding Remarks

The commercialisation and modernisation of the wool industry in China has created a great need for the development of effective industry standards or grading systems for raw wool and wool tops. Chinese wool industry officials have been aware of the requirement to develop a sophisticated grading system to meet the demands of the emerging Chinese wool industry. Pragmatic considerations, however, have seriously constrained the extent to which the recent revisions of the standards could address these requirements. Nevertheless, China now has a remarkably detailed set of raw wool and top standards, given the general level of economic development in the country. While the implementation of these grading systems and other marketing regulations at the grass roots is undoubtedly less than complete, at least there are standards in place.

Furthermore, as will be discussed in Chapter 6, as the required equipment and trained people become available to implement properly more advanced fibre testing and grading systems, the present standards will be further revised. In the meantime, any foreign agribusiness entrepreneur seeking a niche in the Chinese wool industry needs to be aware of the background to, and the deficiencies of, the present industry standards by which wool and wool tops are marketed.

Chapter 5

A Fair Price?

Among the key agribusiness reforms designed to modernise the wool industry and wool marketing is a move to market-determined prices. Commodity prices serve both to allocate resources and to distribute income. Given that the development of the commercial wool industry in China has coincided largely with the formation of the People's Republic, the latter role has been emphasised in the past. Administratively-determined prices sought to determine "fair" prices on income distribution grounds. Where the administrative prices move broadly in line with developments in a particular market, they can be sustained. However, where a divergence arises, both the market of interest and related commodity markets can become seriously out of balance, creating various market distortions and related policy problems.

For China, getting the right balance between income distribution and efficient resource allocation with respect to the wool market is particularly problematic. Herders who derive much of their income from wool are among the poorest groups in the country, and typically it is these poor herders who supply the lower-quality wool. On the other hand, the economic reforms since the late 1970s have heightened the price responsiveness of wool supply and demand, and as a result the administratively-determined prices have become increasingly inappropriate and have led to major wool market imbalances. Weaning market participants off administratively-determined prices, while both avoiding potentially unacceptable implications for the distribution of income and also creating the right market and institutional environment to enable prices to achieve optimal resource allocation, presents major challenges for Chinese policy-makers. Some of these challenges are highlighted in this chapter.

In principle, the agribusiness reforms have involved a move to free markets since 1992, with the price of raw wool being determined by underlying supply and demand factors. The first part of the chapter examines the potential impact of these reforms on the market distortions created by the administratively-determined prices of the past, including distortions across related commodity markets, wool types, wool-producing localities, and wool product markets. Although in principle the wool market was fully deregulated in 1992, in practice the pricing of wool is still strongly influenced by non-market factors. The second part of the chapter, therefore, discusses why a genuinely free market for wool in China is yet to emerge. Both the obstacles to potential new buyers in the wool market and the problems the State is experiencing because local authorities still want to intervene in the market and influence the price of wool are examined. The move to free markets has had an impact on many organisations and especially those institutions traditionally involved in the setting of State prices. The final part of the chapter considers what (useful) new roles these institutions may play in more open markets. In particular, the need for a public market-reporting service is identified. The possibility that the network of price bureaus could redirect their efforts from setting and policing State prices to monitoring and reporting free market prices is canvassed. Foreign market-reporting agencies could play a major role in the modernisation of market information services in relation not only to the wool market but also to the Chinese agribusiness sector more generally.

5.1 Impact of Deregulating the Wool Market on Price-Related Distortions

Relative prices had only a modest role in the allocation of resources in the PRC before about 1980. But as people have been permitted to trade more goods and services in free markets, and households have been given increasing freedom in relation to production decisions, relative prices have become major determinants of how resources are used, when they are used and for what purpose. In this context, major distortions in resource use can arise if the markets for some commodities are opened to free trade while the prices of related commodities remain regulated. This was more-or-less what happened in the case of wool prior to 1992. However, de-regulating the market for wool and wool-based products may not automatically remove the distortions.

5.1.1 Distortions in Related Product Markets

Wool exhibits a range of important relationships with other commodities. Wool competes with other natural fibres and various synthetic fibres in the production of textile products. As one joint product of sheep, wool competes actively for pastoral resources with other pastoral products such as beef (cattle) and cashmere (goats). Cashmere, of course, also competes with wool on the demand side. Furthermore, wool does not exhibit a fixed technical relationship with its joint product mutton, and movements in relative prices create the incentives for herders to raise either fine wool or meat sheep.

The complexity of these economic relationships has exposed the risks in administratively determining wool prices, and has led to significant distortions in related-product markets. To fully assess the effects of past price distortions in the wool market and the impact on them of recent agribusiness reforms, the extent of past distortions and any recent reforms in other commodity markets must also be considered. This section does not attempt to disentangle all the related market distortions. Rather it examines some of the most important developments in relative prices which have occurred as a result of the recent reforms.

Prior to 1956, as pointed out earlier in Section 3.1, the coarse wool produced by nomads was sold during so-called "Nadam Fairs", and prices were not subject to any form of government control. Communications both over time and over space between the small, isolated Nadamur markets were poorly developed. Prices were unstable, unpredictable and inefficient in the sense that they did not reflect any overall interaction of total market supply and total market demand.

With the introduction of a State-controlled unified purchase and distribution system in the pastoral region in 1955/56, the prices paid by the State for quota production of pastoral products were kept artificially low for more than 20 years through the monopoly buying powers vested in the SMC (Table 5.1). Over-quota or residual production, which could be sold on the free market if one existed, was usually too small relative to total output to influence the overall price level.

Table 5.1 Mixed Average Purchasing Prices for Pastoral Products, Selected Years: 1952 to 1992

Year	Raw wool		Raw cashmere		Mutton		Beef	
	Price	Index	Price	Index	Price	Index	Price	Index
	(¥/kg)		(¥/kg)		(¥/kg)		(¥/kg)	
1952	2.08	100	6.98	100	0.52	100	0.43	100
1965	2.18	105	7.60	109	0.94	181	0.85	198
1978	3.40	163	8.20	117	0.87	167	0.88	205
1980	3.43	165	11.11	159	1.28	246	1.34	312
1985	5.04	242	29.94	429	2.33	448	2.91	677
1986	6.00	288	50.80	728	2.39	460	3.00	698
1987	6.29	302	70.17	1,005	2.92	562	3.50	814
1988	10.79	519	128.84	1,846	3.94	758	4.48	1,042
1989	9.07	436	182.69	2,617	4.52	869	5.29	1,230
1990	6.64	319	129.54	1,856	4.12	792	5.19	1,207
1991	6.02	289			4.24	815		
1992	6.96	335			5.19	998		

Source: *Chinese Statistical Yearbook* (various issues), Zhang (1994).

The two decades between the launching of "the Great Leap Forward" in May 1958 and the start of the economic reforms in late 1978 were a period of great political upheaval in China. State prices for pastoral products stagnated during the Great Leap Forward (1958–1961) and free markets largely disappeared. The recognised failure of the Great Leap Forward forced the government, among other things, to raise the purchase prices for animal products in 1961 by 19.3% and once again allow over-quota output to be sold in free markets. However, with the onset of the Cultural Revolution in May 1966, free markets were banned and once again State purchase prices stagnated until 1976 when they were raised slightly.

Despite the dawning of the new era of economic reform after 1978, the market for raw wool remained tightly State-controlled until 1985. Initially, the economic reforms focused on reforming rural production and raising the prices of staple commodities like grains and some meat products (Fig. 5.1). Thus while wool prices rose only slightly in the first half of the 1980s, the prices of agricultural products in general rose more than 50% between 1978 and 1984. As discussed in Section 3.3, it was not until 1985 that the wool market was directly affected by a major economic reform. From 1985 onwards, the provinces were free to establish their own State prices and as a consequence the official State price could vary from one province to the next. The two major wool-growing provinces, IMAR and XUAR, did indeed elect to establish different State prices. Furthermore, both these provinces attempted to prevent the development of a free market for wool during the "wool wars" (Section 3.4).

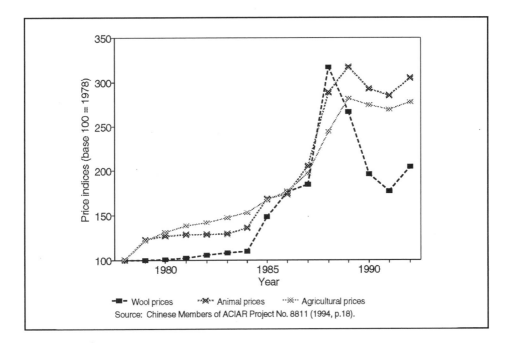

Fig. 5.1 Indices Showing the General Level of Prices for Wool, Agricultural
Products and Animals, 1978 to 1992

The years 1985 to 1991 proved to be a key period for the wool market and wool
pricing. The wool-marketing reforms and subsequent "wool wars" led to a dramatic
rise in wool prices, with average prices in 1988 being more than double those of three
years previously (Table 5.1). Longworth and Williamson (1993, pp.101–105 and
162–164) outline in detail the events that surrounded wool pricing in the major wool-
growing provinces of IMAR and XUAR over this critical 1985 to 1991 period. In
general, however, it is evident from Fig. 5.1 that while other agricultural prices rose
during this period, they did not match the dramatic rise in wool prices. Indeed, by
1988 wool prices had finally caught up with and, in some cases, exceeded the growth
in prices in agricultural commodities since the start of the economic reforms a decade
earlier.

The rapid rise in wool prices altered the price relativities with other agricultural
commodities, especially with other pastoral products with which wool production
competes for resources. A key production relationship or substitution exists between
wool and cashmere, and as already mentioned sheep and goats are often raised in
mixed flocks throughout the pastoral region. (See Longworth and Williamson (1993)
for details.) The composition of these mixed flocks over time has been responsive to
changes in the price relativities. Cashmere prices were even more volatile than wool
prices during the 1980s (Table 5.1). Cashmere marketing was traditionally controlled
by the foreign trade sector rather than the SMCs. Prior to 1985, the fixed State price
for cashmere was very low, especially relative to the prevailing world price. When the
market for cashmere was liberalised in 1985, cashmere prices rose dramatically and
by 1989 were more than 10 times their pre-1985 level. Thus, although wool prices

rose dramatically during the "wool wars", they did not match the rise in cashmere prices and some substitution from sheep to goats occurred. The rapid rise in cashmere prices when combined with the poorly developed cashmere-marketing system in many parts of China led to adulteration of cashmere and downgrading of Chinese cashmere in world markets.

The rapid rise in wool prices also altered the price relativities with competing chemical fibres. Between 1985 and 1990, the amount of chemical fibres blended with wool rose from 49kt to 85kt (Lin, 1993). Almost four-fifths of these fibres were acrylic. As Table 5.2 reveals, the share of chemical fibres in all wool-based products in China rose from around 22% in 1985 to 42% in 1990. Much of the substitution of synthetics for wool in blended products occurred from 1988 to 1990. Furthermore, in 1988 and 1989, the total consumption of chemical fibres by the wool textile industry was 200kt and 211kt respectively or almost three times the amount of chemical fibres blended with wool. Thus almost two-thirds of the productive capacity of the wool textile mills was used to process chemical fibres into purely synthetic fabrics in 1988 and 1989.

Table 5.2 Raw Materials Used to Make Wool-Based Products, 1985 to 1990

Year	Total	Domestic and imported wool*		Chemical fibre	
		Quantity	Share of total	Quantity	Share of total
	(kt)	(kt)	(%)	(kt)	(%)
1985	224	175	78.1	49	21.9
1986	228	163	71.9	64	28.1
1987	252	188	74.6	64	25.4
1988	244	186	76.2	58	23.8
1989	215	146	67.9	69	32.1
1990	200	115	58.0	85	42.0

*Domestic wool is clean wool. Imported wool is raw (greasy) wool and wool top.
Source: *Almanac of Chinese Textile Industry* (various issues).

The move into synthetics by wool textile mills can be directly related to the escalating costs of greasy wool in the late 1980s. The escalating costs were a function not only of higher wool prices (Fig. 5.2) but also of the rapidly deteriorating quality of domestic wool at the time. With access to imported wool severely constrained in 1989, mills turned to the relatively lower-priced chemical fibres rather than the higher-priced, low-quality domestic wool.

As outlined in Chapter 3, by mid-1989 the market conditions for raw wool were much less buoyant than at the beginning of the 1988 season, which had been the peak of the 1980s wool boom in China. In recognition of the depressed market, the IMAR provincial government changed the way the official purchasing prices for the IMAR would be established for 1989. It requested the provincial Price Bureau to set maximum and minimum (reserve) prices rather than a single set of fixed prices. The

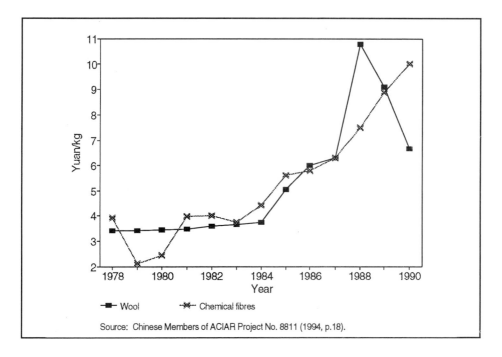

Source: Chinese Members of ACIAR Project No. 8811 (1994, p.18).

Fig. 5.2 Wool and Synthetic Fibre Prices, 1978 to 1990

range of prices was supposed to take into account: the need for an adequate remuneration to herders, the interests of processors and consumers, and the market situation. Reconciling these conflicting objectives proved too difficult for the provincial Price Bureau and, despite the deteriorating market situation, it set maximum official prices in 1990 which were the same as in 1989. Furthermore, it actually raised the minimum (reserve) prices for all types by 27% relative to 1989. As discussed by Lin (1993), this kind of official pricing policy effectively priced domestic raw wool out of the textile fibre market in China in 1989 and 1990, and as already mentioned the mills turned to synthetics since imported wool was not available. The proportion of wool fabric classified as pure wool declined from 33% in 1987 to 15% in 1990 and there were similar drops in the proportion of knitting yarn and blankets made purely from wool. However, as the data in Table 5.3 demonstrate, there were major increases in the amount and the share of wool fabric and wool yarn described as being "pure wool" between 1990 and 1993. The figures in Tables 5.2 and 5.3 provide convincing evidence of the remarkable degree of flexibility in the wool textile industry. Mills clearly have considerable capacity to alter their mix of raw material inputs in response to relative prices and other factors.

The unrealistic official prices in 1989 and the general lack of demand for raw wool in 1990 and 1991 led to the SMCs holding large stocks of wool for which they had paid excessively high prices. Despite financial support from the Agricultural Bank of China (ABC), the ability of the SMC to purchase all wool on offer was in serious doubt in the early 1990s. The situation became so bad that the Central government sanctioned the use of IOUs to enable the SMCs to continue buying wool at the official State prices. At the same time, the SMCs sought to circumvent price controls

Table 5.3 Output of Major Wool Textiles and the Proportion of the Total which was Pure Wool, 1978 to 1993

Year	Wool fabric			Knitting wool			Wool blankets		
		Pure wool fabric			Pure wool yarn			Pure wool blanket	
	Total output	Output	Prop. of total output	Total output	Output	Prop. of total output	Total output	Output	Prop. of total output
	- - - ('000m) - - -		(%)	- - - ('000t) - - -		(%)	- ('000 pieces) -		(%)
1978	88,850	17,140	19.3	37.8	10.3	27.2	6,250	na	na
1979	90,170	22,160	24.6	44.4	11.6	26.1	6,900	na	na
1980	100,950	27,550	27.3	57.3	13.8	24.1	8,840	na	na
1981	113,080	35,260	31.2	76.5	18.6	24.3	10,670	na	na
1982	126,690	46,770	36.9	92.5	17.7	19.1	13,790	na	na
1983	142,910	54,960	38.5	102.1	20.4	20.0	16,220	na	na
1984	180,490	64,490	35.7	110.0	24.0	21.8	17,450	na	na
1985	218,160	82,000	37.6	125.6	27.8	22.1	20,150	4,030	20.0
1986	251,860	74,010	29.4	149.1	26.1	17.5	24,220	4,690	19.4
1987	265,380	87,980	33.2	204.7	36.4	17.8	30,190	5,830	19.3
1988	286,090	85,580	29.9	220.5	48.3	21.5	35,160	7,130	20.3
1989	279,620	58,390	20.9	250.0	41.4	16.5	30,810	5,640	18.3
1990	295,050	44,240	15.0	249.9	30.0	12.6	22,960	2,620	11.4
1991	311,410	55,030	17.7	282.5	47.7	16.9	24,000	2,340	9.8
1992	337,920	73,670	21.8	350.6	85.2	24.3	24,570	2,790	11.4
1993	353,830	89,210	25.2	343.5	112.2	32.7	35,350	na	na

na, Not available.
Source: Chinese Members of ACIAR Project No. 8811 (1994, p.23).

by downgrading the quality of wool offered to it by farmers. Whether this strategy involved deliberate downgrading or simply a stricter application of the purchase standard is a moot point. The data analysed in Section 4.5 support the downgrading hypothesis, at least in relation to 1991. Considerable debate also arose over whether the SMC could accept over-quota wool. In theory, the SMC had a social responsibility to herders to accept all over-quota wool and to pay at least quota-wool prices. In reality, officials admitted at the time that even assuming the over-quota wool was accepted, the price for over-quota wool would be well below that of quota wool.

In light of the problems experienced in the early 1990s and in line with the government's reform agenda and market developments, wool markets were formally deregulated in the major wool-producing provinces in 1992 (Section 3.6). In 1991 in the XUAR, perhaps as a transitional step towards the opening of the market in 1992, prefectural administrations were for the first time allowed some flexibility in the setting of wool prices in their area of jurisdiction. They could vary prices ±10% around the prices set by the provincial Price Bureau. In some cases, even counties were given permission to adjust wool prices in this way. The opening up of the market in 1992 offered scope both for adjustments in the general level of prices and in the differentials across grades.

Although in principle the price of raw wool in China has been determined by supply and demand in the context of a free market since 1992, in practice the pricing of wool is still strongly influenced by other factors. In particular, with SMCs still the dominant buyer in many areas, political as well as economic forces continue to influence wool prices. The following discussion examines some of the more immediate reactions to the opening of the wool market in 1992.

Expectations about how prices would adjust following a re-opening of the market reveal much about how the various participants perceived the change. Prior to the commencement of the purchasing season in 1992, all mill managers in the XUAR met as an informal cartel to decide on prices they would be prepared to offer for the upcoming season. Their agreed price of ¥7.5 per kg for fine wool grade I purchased directly from growers was in sharp contrast with the views of the managers of the provincial SMC who were seeking no change from the 1991 State price of ¥9.70 per kg.

The expectations provided by SMCs in two of the key wool-producing counties immediately prior to the main purchasing season (June 1992) are listed in Table 5.4. In Cabucaer County in XUAR, expectations were for a significant drop from the high 1991 State prices. However, these reflected more the high official prices set in 1991, as actual purchase prices in 1991 would have been lower given the 10% price variation from official prices sanctioned in that year. One especially noteworthy expectation was for a marked decline in the price of improved fine wool grade II, reflecting the extreme difficulties the county SMC had in selling this particular grade of wool in 1991. In contrast to the expectations of the Cabucaer SMC, Wushen County SMC officials expected the 1992 prices for the top two grades of wool to exceed the State prices established for 1991 by the IMAR Price Bureau. At the same time, they expected the prices of the lower-quality grades such as fine Grade II and the improved grades to fall noticeably. Indeed, the premium between fine wool Grade I and fine wool Grade II was expected to be three times larger than that indicated by the quality differentials in the old National (purchase) Standard.

There are at least two possible explanations for the divergence between the expected prices in Cabucaer and Wushen. First, the State prices in Wushen in 1991 were the official IMAR State prices which were slightly lower than the corresponding prices in XUAR to which the Cabucaer expectations relate. A second and more important reason is that the top grades in Wushen would contain some of the best wool grown in China. The Wushen SMC officials clearly expected this wool to attract a premium in the free market.

Since the opening of the market in 1992, it has been difficult to obtain reliable estimates of actual prices paid to growers. However, data for Chifeng City Prefecture in IMAR show that average actual prices for 1992 were lower than the corresponding prices in 1991. The reduction both in absolute and relative terms was greater for the lower grades with a fall of ¥0.40 to ¥0.50 per kg for the fine wool grades and ¥0.60 to ¥0.80 per kg for the lower-priced improved wool grades. Both the actual prices paid in Chifeng City Prefecture and the expected price data (Table 5.4) indicate that the opening of the market in 1992 broke the traditional nexus between grade-price differentials and the quality indices written into the old National (purchase) Standard.

Table 5.4 Wool Price Expectations Just Prior to the Opening of the 1992 Wool-Buying Season

A. Cabucaer County (Yili Prefecture, XUAR)

Type of wool	Grade	Official XUAR Price Bureau purchase prices in 1991	SMC estimates of expected purchase prices in 1992
		- - - - - - - - - (¥/kg greasy) - - - - - - - -	
Special fine		10.38	9.32
Fine	Grade I	9.70	8.79
	Grade II	8.50	7.50
Improved fine	Grade I	6.80	6.80
	Grade II	5.30	3.00
Belly wool		5.00	5.00
Coloured wool		3.00	3.00
Local wool		1.50	1.00 to 2.00
Head/leg/tail wool	(a) Clean	3.00	}1.00 to 2.00
	(b) Dags/wet	1.00	

B. Wushen County (Yikezhao Prefecture, IMAR)

Type of wool	Grade	Official IMAR Price Bureau purchase prices in 1991	SMC estimates of expected purchase prices in 1992
		- - - - - - - - (¥/kg greasy) - - - - - - - -	
Special grade		8.72	8.80
Fine	Grade I	8.02	8.20
	Grade II	7.22	6.60
Improved fine	Grade I	7.16	6.60
Belly/head/leg wool		2.40	2.00

Attempts to obtain reliable data on actual prices paid also uncovered evidence of another possible beneficial effect of the deregulation of the market in 1992. Although it was agreed by county government and SMC officials in both Alukeerqin and Balinyou Counties that the actual prices paid had fallen for all grades of wool in 1992 relative to 1991, the actual weighted average price for all wool purchased by the SMC in both counties increased significantly (by 6.5% and 10% respectively) in 1992 compared with 1991. In fact, it seems that this was the case throughout IMAR. One possible explanation for this apparent paradox is that much more wool of the higher grades was produced in 1992. However, seasonal conditions were not markedly better in 1992 than in 1991 (indeed both were comparatively good wool-growing years) and there is no other possible biological reason for a significant improvement in the average quality of the clips concerned. Another plausible explanation is that the actual (or potential) threat of competition from other buyers, given the free market conditions which existed in 1992, forced the SMC to cease downgrading wool. As mentioned earlier, one of the strategies thought to have been adopted by the SMC to circumvent the requirement that they pay unrealistically high State prices in the 1989 to 1991 period was to deliberately downgrade wool.

When the Chinese government first moved to deregulate the market for wool in 1985, the world wool market was entering a major boom. The international wool boom led to an unprecedented surge in the domestic price of wool in China. Speculation was rife and a chaotic market situation emerged. The general world market for wool was much less buoyant in 1992 when the Chinese authorities once again moved to create a free market for wool within China. The experiences of the 1985 to 1988 period have left their mark. The SMC, for example, has been much more proactive and less reactionary in the post-1992 period than in the late 1980s, and the various authorities concerned with wool are better prepared to manage the transition to a free market in the 1990s than they were in 1985. Nevertheless, it may be some years yet before all participants in the wool market fully accept the wisdom of allowing prices to be determined entirely by market forces.

5.1.2 Distortions across Types of Wool

Both when the State prices for wool were established by the National Price Bureau (before 1985) and also when State prices were set by provincial price bureaus (from 1985 to 1991 inclusive), premiums and discounts for the various grades relative to the reference grade were defined by the quality differentials in the old National (purchase) Standard (Table 5.5).

Table 5.5 Quality Differentials in the "Old" National Wool Grading Standard

Type of wool	Grade	Quality differential
		(%)
Fine and semi-fine	Special Grade	124
	Grade I	114
	Grade II	107
Improved fine and semi-fine	Grade I	100
	Grade II	91

Source: Appendix A.

Improved fine wool Grade I was the reference grade throughout China until 1986. In that year, XUAR elected to use fine wool Grade I as the base grade but there was no change elsewhere. Once the State price for the reference or base grade had been decided upon, the quality indices written into the National Wool Grading Standard in 1976 were applied to determine the State prices for the other grades. The differentials in the State prices for the various grades were, therefore, constant in percentage terms from 1976 to 1991.

Both the need to set State prices according to the grades established by the National (purchasing) Standard and the requirement that grade-price differentials comply with the quality indices incorporated in the standard, confirm the traditional key role of the National Standard in raw-wool marketing in China.

One particularly important aspect of this approach to the setting of State prices was that the official prices for fine wool and semi-fine wool were the same, since the quality indices for these wool types were equal. As fine wool has an average fibre diameter of 25μm or less while semi-fine wool has an average fibre diameter of from 25.1μm to well over 40μm, these two wool types are suitable for entirely different end-uses and, as a consequence, can be expected to have different values. Normally, fine wool would be expected to be the more valuable. As the State prices were established on a greasy (raw) wool basis, setting the same State price for fine and semi-fine wool usually did result in a significantly higher price being paid for the fine wool on a clean basis because the semi-fine wool yielded a higher percentage of clean wool than the fine wool. However, the premium paid for fine wool on a clean basis was arbitrarily determined by the yield of any particular lot of wool rather than by any consideration of real value.

From the viewpoint of wool growers who are paid on a greasy-wool basis, the State prices traditionally offered were the same for fine and semi-fine wool. There appeared to be no price incentive to upgrade towards finer wool. In fact, given the extra costs (and care) needed to raise fine wool sheep, the State pricing policies actually discouraged the raising of fine as opposed to semi-fine wool sheep. Once the widespread adoption of the household production responsibility system in the early 1980s gave individual households greater freedom to respond to market incentives, the traditional State pricing system became a major factor inhibiting the upgrading of private flocks towards finer-woolled sheep.

Clearly, State raw-wool pricing policy was inconsistent with the emphasis being placed on breeding programs and other measures aimed at encouraging a swing to finer-woolled sheep.

The traditional State pricing system was developed for a centrally planned economy where prices determined incomes but did not directly influence what was produced. That is, the fixed grade-price differentials for wool and the associated failure to differentiate between fine and semi-fine wool prices evolved to address income distribution problems at a time when production decisions were not influenced by price incentives or disincentives. Reforms in relation to how rural production was organised in the early 1980s, which gave individual production units the opportunity to respond to price incentives, created an urgent need for market reforms which would allow prices to reflect true social values more accurately. However, in the case of raw wool, a move towards more appropriate differentials between grades was considered likely to have a profound impact on the distribution of income in wool-growing localities. Consequently, it was not until 1993 that the authorities were prepared to abolish formally the concept of official grade-price differentials for wool.

Given the large stochastic shocks to Chinese wool supply and demand through time, fixed, administratively-determined grade-price relativities are poorly suited to changing conditions and have resulted in significant market imbalances among the different types of wool at particular points in time. This is highly likely to be the case even where price administrators have significant information resources and the ability to change grade-price differentials regularly (as demonstrated by the ill-fated wool reserve price scheme in Australia and various EC Common Agricultural Policy price support regimes). Limited information flows and the fixed nature of the differentials for Chinese wool, however, resulted in major market imbalances and the build-up of large stocks in some of the lower grades in 1989 to 1991. As pointed out previously, this led to purchasing agents such as the SMC adopting other strategies such as deliberate downgrading to create more realistic grade-price differentials.

With the widespread deregulation of wool markets in 1992, Chinese wool-marketing institutions are gradually moving away from the system of administratively-determined and fixed grade-price differentials. A properly functioning free market system would automatically generate the appropriate premiums and discounts for the various types of wool and would eliminate the need for the creation of arbitrary, *de facto*, grade-price differentials via incorrect grading. However, given the traditional inertia in China, it will be some time before properly functioning markets can deliver the appropriate grade-price differentials at all levels in the marketing chain.

Even if a properly functioning free market can generate the appropriate grade-price differentials, this may not be sufficient to relay the preferences of mills (and ultimately of the consumers) to herders. For example, with more open markets in fabrics and garments, mills now receive different prices for products made from wool of different mean fibre diameter. While mills may be prepared to pay premiums or discounts for wool of different mean fibre diameter, wool graded according to the old National Wool Grading Standards was simply too variable in regard to this quality characteristic. That is, the traditional purchase standards were too imprecise to allow the marketplace to properly value certain important wool quality determinants such as average fibre diameter.

The new National Grading Standard attempts to address this problem and, to the extent that it is correctly applied, it may enable clearer price signals as regards the relative value of the various types of wool to reach producers. As explained in Section 4.3, the new Grading Standard not only provides for wool to be graded by average fibre diameter but it also decouples the pricing of fine and semi-fine wool much more clearly than in the old National Standard.

5.1.3 Distortions across Different Locations

Past pricing arrangements have not only cross-subsidised different types of wool but also wool produced at different locations. Although wool prices have differed between provinces, within the province there was a single State price for each type of wool. That is, the official prices reported and discussed in this chapter for any one province were prices which were supposed to be paid to the herders at all purchase stations irrespective of the location of those purchase stations within the province in question. The costs of transporting wool from the grass-roots SMC purchase station to the higher-level SMCs was absorbed by the higher-level SMCs. The equalisation of farm-gate, or at least purchase-station, prices has meant that wool produced or supplied in remote regions has been cross-subsidised by wool grown in more favourably located areas. Given the vast areas over which wool is produced and the poor transport infrastructure, the extent of this cross-subsidisation must have been substantial in some instances. (Transport costs for greasy wool in 1992 were around ¥0.30 per tonne per km.)

The deregulation of wool markets should have a major impact on the extent of this freight equalisation and may serve to integrate the wool markets better in a spatial sense. More favourably located producers could expect to receive a higher relative farm-gate price than their more remote colleagues. The definition of favourable location depends on a number of factors including the degree of competition for wool supplies in a given area. For instance, it is not difficult to envisage a herder who is more distant from one mill, though closer to a neighbouring mill, receiving a higher farm-gate price than a grower who is closer to the first mill because of different degrees of market competition in the two localities concerned.

Of course, to the extent that certain grades of wool are produced in certain localities, there may be considerable interaction between grade-price differentials and location-price differentials. As a result of differences in transport costs and other location-specific factors, the premiums received for a particular grade relative to some reference grade may vary considerably across the pastoral region. These market-determined differences in farm-gate prices may take some explaining in China where the traditional State pricing system virtually ignored locational differentials and inappropriately allowed for grade differentials.

The opening of wool markets throughout China in 1992 may also limit the spatial price distortions which arose following the wool-marketing reforms in 1985 in which the various regional governments pursued different open or closed market policies. During the 1985 to 1988 period, despite the fixed prices set by the IMAR Price Bureau, purchase prices paid to farmers by the SMC in those areas where the market remained officially closed varied widely from prefecture to prefecture (and even between counties within the same prefecture). There were two main reasons for this situation. First, some prefectures (and counties) wanted to encourage wool production and were not concerned about the cost of the raw material to textile mills. The prefectural (and county) price bureaus in these areas tried to set the official purchase prices as high as possible. On the other hand, local governments (at the prefectural and/or county level) which owned and operated textile factories were keen to keep raw-wool prices down to increase the profitability of their factories. The second reason for the growing regional variation in official raw-wool prices during the 1985 to 1988 period was the emergence of the "black market" for wool in these officially closed areas, with many units and individuals competing with the SMC for the available wool. This "private" illegal competition was especially important in counties (and prefectures) adjoining provinces such as Liaoning where a vigorous free market in wool developed from 1985 onwards.

5.1.4 Distortions in Wool Product Markets

As with the pricing of raw wool, official policy has moved rapidly towards allowing a free market to exist for wool products in the 1990s. But again the shift away from administered pricing has highlighted the interdependence and potential conflict between various policy objectives relating to industry efficiency, income distribution, and consumer sovereignty. An understanding of traditional practices in relation to wool product pricing is an essential starting point for a meaningful appreciation of likely future developments.

The prices of textiles, fabrics and garments were determined administratively from 1951. Although there were some notable amendments to the system in 1966, 1978 and

1986, regulated prices were essentially based on cost of production. For example, the prices of wool fabrics at the factory gate were set to cover average costs, a profit margin and to allow for a tax on wool fabrics. Costs included raw-material costs, dyeing costs, packing costs and labour and miscellaneous costs. Dyeing costs varied and were related to the type of material dyed (such as tops, clean wool or unfinished wool fabrics) while packing costs were based on standards set out for packaging. Labour and miscellaneous costs included all residual costs such as power, fuel, salaries, maintenance and management. Profits were then imputed as a proportion of the input costs. Separate profits were imputed for labour and miscellaneous costs and for all other costs. Although the tax rate was assumed to be 14.5% on all products, in provinces such as Gansu, the tax rate could be varied according to the state of the market and according to any profit squeeze being experienced by processors.

Mark-up pricing based on cost of the raw materials plus the cost of processing established a direct link between official raw-wool and wool product prices. This created some special problems for mills in the late 1980s. The rapidly rising raw-wool purchase prices during the "wool wars" were combined with declining quality and clean yields. Clearly, the mills faced a major escalation in costs. Wool product prices rose dramatically, with the price for a typical woollen blanket, for instance, rising from ¥65 in 1984 to ¥190 in 1988. However, the magnitude and rapidity of the implicit rises in clean-wool prices meant that they could not immediately be incorporated into the wool product prices, especially as product prices negotiated with the price bureaus were normally set for the whole season. Processor margins were squeezed and many mills incurred losses. Using various simplifying assumptions, Lin (1993) estimated that mills incurred losses of around ¥3 per m on fabrics and ¥20 per kg on hand-knitting wool yarn.

In an attempt to reduce their losses, many textile mills sold their products at much higher prices than the official wholesale State price, and consequently retail prices were also pushed up and consumers reacted by cutting back their purchases of wool-based goods. In XUAR, the provincial government responded by deregulating the fabric and garment market. In 1989, the IMAR government reduced the purchase prices for raw wool and raised the wholesale prices for wool-based fabric. The State wholesale price for worsted goods reached ¥40.70 per m and for woollen yarn it was ¥5,683 per 100kg. On the basis of these prices, the mills could operate profitably but retail prices were pushed even higher, consumer resistance increased still further, and even fewer wool-based goods were sold. As a result, the mills purchased only about half as much raw wool from the SMC in 1989 as they did in 1988 and, as already mentioned, the SMC stockpile grew to be very large. The high prices textile mills were forced to pay for raw wool reduced consumer demand and helped create the domestic raw-wool surplus problem in the 1989 to 1991 period.

Ultimately, more open wool and wool product markets should avoid the problems that arose in the pursuit of objectives in the regulated markets and that were inconsistent with the underlying supply and demand conditions in these markets. However, the immediate effect of opening the wool product markets in 1992 was to squeeze mill returns further as fabric prices fell while mills still had to use their high-priced wool stocks.

5.2 "Open" Markets versus "Free" Markets

In China, a Central government decision to open a market does not guarantee that completely free trade will occur throughout the country. For example, the Central government "opened" the market for wool in 1985. A free market for wool quickly emerged in some areas while provincial and/or local governments in other areas vigorously resisted free trading. The attempt to develop an open market for wool was aborted in 1989 when the State recentralised control over the purchasing of raw wool. In 1992, it was decided to try again and once more the wool market was declared open, a decision agreed to by all the relevant provincial governments.

In principle, beginning in 1992 it should be possible for a completely free market for raw wool to develop throughout China. In practice, however, completely free trading in raw wool will take some time to emerge. Not all entities wishing to purchase wool are being treated the same way by local authorities in wool-growing areas. Furthermore, some local governments continue to intervene directly in wool pricing in various subtle ways.

5.2.1 Preferential Treatment for SMCs

The first obstacle to free trade in wool is the threat such a change poses for the economic viability of the SMCs. In the past, the SMCs have applied a fixed mark-up to cover the costs of marketing. According to the regulations, raw wool was purchased by the grass-roots SMC at the township level and then shipped to the county-level SMC which in turn "sent" it to the prefectural SMC. In reality, often the wool was not handled or even sighted by the prefectural-level organisation but was shipped directly to the mills from the county. The prefectural SMC had the right to sell the wool to the textile mills. The fixed marketing margin between the State purchase price paid to growers and the price at which the wool was sold to the mills was distributed among the different levels of the SMC.

The percentage mark-up varied across provinces/regions as did the precise breakdown among the different levels in the SMC. The usual margin in the IMAR was 27%, of which 10% was the product tax and a typical breakdown of the remaining 17% among the SMC levels was 9% for the township SMC, 4% for the county SMC and 4% for the prefectural SMC. However, the distribution of the marketing margin could vary depending on the locality. For example, in Wushen County, of the usual 27% margin, 23% went to the township SMC (of which 10% was the product tax which was paid to the county government) and 4% to the county SMC, with no share going to the prefectural SMC. Furthermore, agreement was reached that if the county Animal By-Products Company could sell the wool with an even bigger margin, then the extra profit could be shared with the grass-roots (i.e. township) SMC. Note that under this agreement the profits would not be shared by herders even though theoretically they own the grass-roots SMC. This arrangement was known as the "separate purchase–joint sale" system.

In Gansu Province, the 26.33% marketing margin consisted of a 10.1% wool tax, a 5.4% commission for the grass-roots SMC, 4.03% for the county-level SMC, and 6.8% to the provincial-level SMC, with the prefectural SMC having no role in wool marketing.

The non-wool tax component of the marketing margin has been critically important to the SMCs. The margin covered costs of transport, packaging, grading, sometimes shearing, interest, administration salaries and management fees for SMC staff and pensions for retired employees. The traditional 17% (in the IMAR) marketing margin paid to the SMCs was established as far back as 1963, and developments in the late 1980s and early 1990s were placing great pressures on that margin. In recent years in the pastoral region, the SMCs have both increased their staff and been forced to support a growing number of retired staff. The latter responsibility has placed a considerable burden on SMCs as pensions are increased in line with salary levels and little provision has been made in these organisations to cover future pension requirements. At the same time, the SMCs faced increased interest charges and increased costs of other marketing inputs. For instance, interest rates of around 5% in 1963 had increased to 13% in the early 1990s. Consequently, during the wool crisis in 1989 to 1991 the SMC organisation was arguing for an increase in the percentage mark-up on wool. Indeed, senior staff at the provincial level of the SMC in the IMAR were claiming that their share of the mark-up should be at least doubled from 17% to 34% or should be even 38% of the purchase price paid to farmers.

The representatives of local governments and the SMCs argue strongly in favour of the old fixed marketing margin between farm-gate and mill-door prices. They point to the social benefits of having a viable local SMC organisation not only to buy wool and other products but also to supply the necessities of daily life to the people. In addition, the SMC is a major source of employment in many pastoral townships. The wool tax enables the county government to provide better services and infrastructure and, as with the SMC, the county governments are major employers of local people.

A genuine free market is perceived as a serious threat to the economic viability of many township and even county SMCs in the pastoral region. Other buyers without the local social obligations of the SMCs can operate on smaller marketing margins and hence they can outbid the SMCs for supplies of raw wool. While such developments may enhance the economic efficiency of wool marketing, it could destabilise pastoral communities socially and politically. For these reasons and because a completely free market also poses a major threat to the fiscal base of many county governments, these authorities are likely to use sometimes rather subtle but nonetheless effective measures to protect the SMCs. That is, while the central and provincial governments insist that policies that permit the free marketing of wool are in place, at the county and lower levels circumstances may be entirely different. That is, in many parts of the pastoral region the existence of a free market for wool is largely what Longworth and Williamson (1993, p.321) have termed a "policy mirage".

5.2.2 Collection of Wool Taxes

The second major obstacle to be overcome in a move to a completely deregulated wool market is the difficulty in collecting wool taxes from a multiplicity of buyers in the free market. A tax of 10% is levied on the value of all raw-wool purchases throughout China. Among the four natural fibres (i.e. cotton, silk, wool and fibre crops), only wool is taxed in this way. Furthermore, meat products (i.e. pork, beef and mutton) are only taxed at 3%. As pointed out in the previous section, the tax was traditionally included in the SMC marketing margin. Under the Tax Regulations passed by the State Council in 1984, when a farmer delivers or sells an agricultural, animal or fish product which is subject to a product tax, it is the responsibility of the unit taking delivery or purchasing the product, traditionally the SMC in the case of wool, to pay the product tax. Note also that when the State purchase price to be paid by the SMC was increased sharply, as between 1985 and 1988, the absolute amount of the tax paid per unit of wool also increased sharply, even though the tax rate remained constant at 10%.

The tax was introduced in 1950 as one means of taxing animal production. In the early 1950s, the nomadic nature of animal production caused difficulties in levying an animal tax on herders. The wool product tax is payable to the county-level government by the wool purchaser prior to sale to the processor. In the case of the SMC, the tax is payable via the county-level SMC to the county government. Free market buyers, in theory, must also pay the 10% wool tax to the county government. In practice, however, it is very difficult to collect the tax from free market buyers. The government has attempted to counter the loss in revenue arising from the difficulty in collecting the tax by insisting that free market buyers pay a tax known as the commercial management fee in addition to the wool tax. This tax or fee is levied by the Industry and Commercial Management Division within the county-level government, and is universally applied throughout China. In addition, free market buyers of wool are required to be registered by the Division of Industry and Commercial Management at the county-government level. Registration, however, does not necessarily guarantee that accurate purchase records are submitted. In any event, it may be extremely difficult to enforce registration in some areas. Indeed, the additional commercial management fee may give extra incentive for the free market buyers to avoid registration and payment of the taxes.

Until 1992, formal permission was required to take wool out of Xinjiang. That is, buyers of wool even if they bought from the SMC were required to have special permission to export wool to other provinces. There was also a 20% export tax on the value of the wool, payable before permission to export wool was granted. This tax was called a pasture construction tax and it was intended that the revenue collected would be used to improve pastures in XUAR. Since there is only one practical route out of the XUAR through Gansu Province, it was easy for the government to police these limitations on the export of wool. In 1992, the export tax was lifted and no special permission is required to take wool out of the XUAR. All buyers of wool, however, are still liable for the payment of the 10% product tax to the county governments.

The wool product tax is a major source of fiscal revenue for local governments in the pastoral region. For example, in Aohan County in Chifeng City Prefecture of IMAR, the wool tax accounted for 10% of the county's revenue even though Aohan is normally considered an agricultural county (Longworth and Williamson, 1993, pp.199–210). In Sunan County, which is a major wool-growing county in Gansu Province, the wool tax raised 25% of fiscal revenues. In many counties, the product tax on wool is a major means of raising the funds to pay the wages of county officials (cadres), health workers and teachers. Since the introduction of the fiscal reforms which began in 1980 and which called for a greater degree of self-financing, local governments have become hard-pressed to balance their budgets. Therefore, modifying the regulations regarding the wool product tax would be a very important change which could have an impact on the living standards of many local government employees and/or affect the rate of infrastructure development in the pastoral region financed by local government.

Freeing-up wool markets and allowing other buyers to compete with the SMC clearly create major problems for county governments in the collection of the wool tax. Although the immediate threat to the SMCs in more open wool markets comes primarily from other government agencies purchasing wool, experience during the "wool wars" (1985 to 1988) showed that it was not as easy to collect the wool product tax from other government agencies and private buyers as from the SMCs which had close links with the county governments. Indeed, the determination of county governments to protect their tax base by preventing free trade in wool to ensure collection of the wool tax is one element often ignored in discussions of the "wool wars".

5.2.3 Income Distribution and Intervention Buying

The third major obstacle to a fully deregulated raw-wool market centres on the political sensitivity of wool prices. The income distributional effects of changing the price differentials which have traditionally existed in the administered pricing schedules for wool are seen as a major barrier to the State completely withdrawing from the price-setting or price-influencing business.

While premiums for finer, better-quality wool would in the longer run encourage farmers to produce more of the "right" kind of wool from the viewpoint of the textile mills (and ultimately the final consumers), in the immediate future premiums could only be achieved by substantially reducing the prices paid for semi-fine and coarse wool. At present, most of the finer, better-quality wool is produced by State farms and better-off farmers with larger flocks. Furthermore, the production of this better wool is concentrated in a relatively few counties (or wool production bases) within the pastoral region. Consequently, the overwhelming majority of farmers raising sheep, of SMCs which purchase wool, and of county governments (which depend on the 10% wool product tax for fiscal revenue) have traditionally been strongly opposed to a wool-pricing system which creates more appropriate premiums and discounts for quality. This has been a major political obstacle to the development of a more appropriate pattern of prices for wool of different types and grades in China. In fact, it is probably the major underlying reason why the free market experiment which began in 1985 was so vigorously opposed in some areas and why there was so much resistance to the development of wool auctions in the late 1980s and early 1990s.

5.3 Changing Roles for the Price-Setting Institutions

The traditional role of price bureaus was not only to set prices for wool and other commodities but to see that they were enforced. Consider, first, the price-setting role. In the IMAR, the provincial Price Bureau under the general guidance of the National Price Bureau and in consultation with the State-owned textile mills, the IMAR Bureau of Textiles, the IMAR Bureau of Animal Husbandry and other interested groups, had the responsibility for setting the official State purchasing prices for wool in the IMAR. In principle, the prices were based on cost and expenditure data and on market outlook information. Surveys were undertaken to examine the costs of producing wool and other pastoral products such as cashmere. The surveys were conducted through the county-level price bureaus. For instance, in Balinyou County in Chifeng City Prefecture, the county Price Bureau conducted surveys on production costs using its own information network down to the village level. Information from the cost surveys was passed on to the prefectural Price Bureau in Chifeng City which then reported the cost survey information to the provincial Price Bureau in Huhehot. After taking into account the market situation, the administratively determined prices were then relayed back along this information chain.

While the formal mechanism for setting prices outlined above would seem to be relatively straightforward, in reality the situation has often been rather different. For instance, the IMAR Price Bureau was only established as a separate institution in 1981. Prior to that date, a number of different governmental institutions had, at different times, the task of setting State prices for the IMAR. The political sensitivity of prices under a centrally-planned system ensured a continual struggle for the right to set prices. The situation was the same throughout China. Given the historical background, therefore, the setting of the State price for raw wool in IMAR in the early 1980s was based on political rather than economic circumstances. This conclusion is supported by the fact that wool prices remained relatively constant from 1981 to 1984 while the prices of many other agricultural and pastoral products increased by up to 50% (see Section 5.1.1 and Fig. 5.1).

Economic forces began to have an impact on State prices for wool in IMAR from 1985 onwards but politics still dominated the final decision. As outlined in the earlier section, State prices lagged behind free market prices until 1989 when they were set at unrealistically high levels. In 1990 and 1991, State prices were reduced but again not by as much as market realities would have suggested. Furthermore, although officially the prices paid to farmers by the grass-roots SMCs were supposed to be the same throughout the wool-buying season (June to October), in the post-1985 era many grass-roots SMCs had begun to vary their prices during the season so that they could compete with other buyers. In the 1985 to 1988 seasons, most growers who stored their wool after shearing were able to obtain higher prices later in the season. However, because the market became progressively more depressed as the 1989 season advanced, the reverse was the case in 1989. State pricing for raw wool in IMAR in the 1985 to 1991 period demonstrated in the Chinese context the well-known difficulties of attempting to set State prices when a parallel free market price-setting mechanism exists.

In the XUAR, as in the IMAR, before a provincial price bureau was established in 1980 a number of different institutions had, at different times, the power to set wool prices. After 1980, the provincial Price Bureau in XUAR set the price of wool in much the same way as in IMAR and with similar consequences.

The other traditional role of price bureaus was to ensure that the prices set were enforced. In particular, county price bureaus were sometimes called price inspection bureaus (or institutes) to emphasise their inspectorial roles. In principle, therefore, the county price bureaus were supposed to ensure that the SMCs paid wool growers the appropriate State price for their wool. The county price bureaus were supposed to inspect the receipts given to herders and record the qualities and quantities of wool purchased by the SMCs. These details could then be compared with the qualities and quantities being sold by the SMCs to the mills. In principle, therefore, the price bureaus were supposed to prevent the SMCs using their monopoly buying power to exploit wool-growing households by downgrading their wool at the time of purchase.

With the elimination of State wool prices from 1992, the price bureaus will retain their monitoring (inspecting) role. However, the essentially private-treaty free markets that have emerged for wool and other products create new opportunities for the price bureaus. At least in the initial formation of these free markets, information flows will be poor. A lack of information could lead to chaotic market conditions and poor integration of markets across time, space and form. In general, well-developed and properly functioning commodity markets elsewhere in the world rely on sophisticated and extensive information networks. In time, the price bureaus are likely to become a State market-reporting organisation providing much-needed information about prices and quantities being traded in the free markets. To do so, however, will require major changes to the way they have operated in the past. No longer can information be coerced from a single State purchasing agency (i.e. from the SMCs). Data collection will be made much more difficult by the multitude of market participants each of which is seeking a competitive edge in terms of the market knowledge it possesses. Furthermore, in the new marketing environment, the types of information required by these agents from a market-reporting service such as a revamped price bureau, will differ greatly from the information the bureaus provided in the past. Consequently, the price bureaus will need to overhaul radically their data-gathering and reporting procedures which evolved under the old regulated marketing arrangements if they are to provide a useful market information service in the future. The magnitude of this task creates opportunities for market-reporting agencies from outside China to assist the price bureaus modernise their operations.

Chapter 6

Testing Times

Central to many of the wool-marketing reforms and to the modernisation of the wool industry in China is the development of appropriate quality standards and grades to facilitate exchange between buyers and sellers. Standards such as those discussed in Chapter 4, however, are of little value unless procedures exist to determine accurately the various wool attributes used in the standards. It is also important to have institutions which are widely acknowledged as independent authorities both to inspect the wool and certify the grading and to settle disputes. This chapter briefly outlines the system of fibre inspection and associated activities in China as they relate to wool marketing.

The need for an adequate system of fibre inspection if the agribusiness reforms are to be built upon and if the wool industry in China is to develop further poses some major challenges. In general, the existing resources devoted to fibre inspection, especially in the main wool-growing areas, are both inadequate and outdated. Few resources are available for re-equipment or for the upgrading of skills. The paucity of resources for fibre testing is compounded by a lack of coordination among various government institutes and State enterprises concerned with fibre testing. Problems in developing an adequate system of fibre inspection are also compounded by the wool production and marketing systems in China which necessitate the testing of a very large number of small, heterogeneous lots of wool in many remote localities dispersed over a vast area.

6.1 Overview of Fibre Inspection and Testing

Wool inspection and sorting occurs at various points in the wool-marketing chain. Traditionally, the first inspection occurred at the grass-roots purchase station where buyers from the township-level SMCs subjectively (by visual and touch methods) appraised and sorted wool delivered by herders. At this stage, the wool was supposed to be graded according to the National Wool Grading Standard to determine the basis for payment to herders. The buyers were expected to compare the wool with standard samples exhibited at each purchasing station before deciding on the grade. The buyer was responsible for separating the belly wool, head and leg wool and wet (daggy) wool before weighing the fleece.

As the quality incentives herders face derive from grading at the grass-roots level, the initial grading has been and will continue to be a vital link in the whole marketing chain. Unfortunately, however, there often have been major problems with this initial inspection and grading. Not only has the wool been poorly graded but also buyers have not always undertaken even the most elementary sorting, as illustrated by the following observations of one of the authors at a purchasing station operated by the Tuke township SMC in Wushen County of IMAR in 1992. The wool arrived in plastic sacks (about the size of wheat sacks) delivered by the herders by donkey cart, motor bikes and small tractor-drawn trailers. Each delivery was of a small volume. The wool was sorted, graded and priced by the SMC buyer operating the purchase station. Upon inspection of an open sack/bale of purchased wool labelled "fine wool Grade I", the wool was found to be unskirted. The fleece wool in the bale was typical dusty, tender 64s count wool, but it also included weather-damaged backs, sweat- and grease-stained inside thigh wool, and wet and daggy rear wool. No belly wool could be found. Clearly, the buyer at this station was not removing the damaged, dirty, wet or otherwise undesirable parts of the fleece before putting it in the bale for dispatch to the next step in the chain.

The grass-roots or township SMCs purchase the wool from the farmers at the purchasing stations and then hand it over to the SMC at the county level. Wool is transported from the grass-roots level to the county level in synthetic or jute fabric bags. Despite the potential for significant foreign-fibre contamination, the wool is not machine-pressed at this stage so contamination is not really a problem. At the county level, the wool is removed from the synthetic or jute bags and machine-pressed into the standard cotton bags or bales used for transporting wool to the mills in China. (The dimensions of these bales are 1m in length, 60cm in width and 40cm in height.) Sometimes the wool may be re-sorted at the county level but, in the past, this has not been common practice.

At the county level, the wool business of the SMC is handled by specialised subsidiaries often called animal by-products companies. As explained in more detail in Section 7.2.2, these companies are responsible for selling the wool to textile mills and other agencies.

Problems with the initial grading at the time of purchase and the lack of any subsequent re-sorting by the purchasing agency mean that virtually all of the wool passing through the SMC system and destined to be made into tops needs to be re-sorted at the mills. Consequently, the older mills have a great deal of experience in wool inspection, sorting and grading. In recent years, some of the mills have been at the forefront in the development and application of new testing techniques for wool in China. Apart from the testing and grading of raw wool, mills have also had a long involvement in testing and grading scoured wool and wool tops and have also been a source of training in wool sorting for employees of the SMC.

Development and commercialisation of the wool industry necessitated the establishment of a separate independent fibre inspection organisation which is known as the Fibre Inspection Bureau (FIB). The FIBs at the various administrative levels now undertake and supervise testing of wool at various points in the wool-marketing chain. Many of the recent wool-marketing reforms imply a greatly expanded role for the FIB, and it has become the primary vehicle for extending the scope of objective measurement and testing of wool along with other fibre inspection procedures.

Another set of institutions engaged in fibre testing are the textile research institutes (TRIs). The role of the FIBs and TRIs in relation to wool fibre testing vary from province to province. As parts of the remainder of this chapter will demonstrate, there is an urgent need to rationalise the activities of these two sets of institutions, given the scarce resources available for the modernisation of fibre inspection and testing in China.

6.2 Fibre Inspection Institutions

While the TRIs and the FIBs are both involved in fibre testing, the principal activities
of the TRIs relate to research and development and will be discussed in Section 6.6.
The organisation primarily responsible for the inspection and testing of domestically-
produced fibres such as wool is the FIB. While another agency, the Commodity
Inspection Bureau, is assigned the task of testing imported wool, the FIB undertakes
the technical aspects of the testing on behalf of the Commodity Inspection Bureau.

6.2.1 Some Background on the FIB

The history of the FIB extends back to 1950 when it was first set up within MOTI as
a quality control unit for Chinese textiles. The National FIB in Beijing is the highest-
level organisation concerned with fibre testing in China. It establishes grading
standards, carries out testing, and acts as an arbitrator to settle disputes about quality
aspects of wool sales. However, it is not concerned with the management of wool
marketing. The National FIB is one of a number of bureaus within the National
Standards Commission under the State Council. (The National Standards Commission
ranks above many Ministries in the Chinese Government.) The National FIB is at the
top of a vertically-administered structure of FIBs with branches down to the county
level in some areas.

The National FIB aims to introduce as quickly as possible the measurement of
three wool characteristics by machine: clean yield, fibre diameter and length. Other
characteristics such as style, crimp definition, colour, grease and suint content, and
fibre strength will continue to be assessed by visual and manual methods.
Surprisingly, despite the high proportion of tender wool in the Chinese clip and the
importance of fibre strength in determining wool quality and mill efficiency, the FIB
has not attempted to establish standards for fibre tensile strength. Apparently the FIB
is reluctant to tackle the issue of fibre tensile strength since it is only obliquely
referred to in the new National Standards introduced in December 1993. (See Section
4.12 in Appendix B.)

In 1990, the FIB issued certificates for 50,000 tonne of wool, of which 21,000
tonne was measured by machine. These amounts represented roughly 20% and 8%
respectively of all wool produced in China in that year. The remaining certified wool
was appraised by manual/visual methods, usually at the county level. Despite the
seemingly low proportion of wool that is objectively measured, it represents a
dramatic increase on previous years. Prior to 1990, most of the wool was appraised by
manual and visual methods, with machine measurement only being used in the case of
disputes. The extended scope of objective measurement has been the direct result of
pressure from the State Council and the State Planning Commission of China. In
1991, it was decreed that all wool being sold to State textile manufacturing firms
should be tested by the FIB. The State Council also decreed that all wool tested by
the FIB should be measured by machine. However, owing to the lack of equipment
and personnel, at the end of 1990 the FIB expected to be able to issue certificates for
only 68,000 tonne of raw wool in 1991 (about 27% of national output), of which only
50,000 tonne was likely to be machine tested.

To gain an appreciation of the type of wool being tested by the FIB, consider the
source of the 50,000 tonne tested in 1990. Roughly half came from XUAR and half
from IMAR. Most of the XUAR wool which was tested was produced in Yili and

Tacheng Prefectures, but there were significant quantities from several other prefectures, especially Shihezi, where most of the wool tested came from State farms. Other than in Shihezi Prefecture, 80% or more of the wool which was tested prior to sale originated from private herders, while the remaining 20% came from State farms. The FIB tested the wool for the county-level SMCs which had obtained the wool from the grass-roots township SMCs, which in turn had bought the wool from herders on the basis of the old National Wool Grading Standard and State pricing system. The wool from IMAR which was tested came mainly from the major fine wool-growing prefectures of Xilinguole and Chifeng City, although 30% of the 2,000 tonne of "Eerduosi" fine wool produced and sold by Wushen County in Yikezhao Prefecture was also tested in 1990. Less than 10% of the wool tested in the IMAR originated from State farms.

Since the FIB only tests wool for the county- or higher-level SMC organisations, the township SMCs have the opportunity to act as "wool merchants" by interlotting, re-sorting and regrading wool of similar types from different flocks or purchasing stations. However, prior to 1992 it seems few if any grass-roots SMCs had taken advantage of this "value-adding" opportunity.

The National FIB personnel are responsible for training FIB staff at the provincial level in fibre inspection technology and procedures. As will be discussed in more detail below in Section 6.3.1, the provincial personnel are then expected to instruct staff at prefectural and county levels.

6.2.2 Fibre Inspection in the Wool-Growing Provinces

The FIB has branches in all provinces which produce significant amounts of natural fibres. In the case of wool, the activities of the FIB in IMAR, XUAR and Gansu serve to illustrate the contribution of the FIB to wool marketing.

IMAR
In IMAR, the FIB is part of the IMAR Bureau of Technology Monitoring and its primary roles are to administer standards for cotton, wool and silk fibres and to provide technical services in relation to the marketing of these fibres. In particular, the IMAR FIB is responsible for:
- Setting up and enforcing standards for fibres. In the IMAR, the FIB applies the national standards for wool, though it has power to vary these if it wishes. In principle, in China there are four levels at which standards might be set: national, regional, professional (industry) and enterprise level. In fact, in the case of wool, the FIB at all levels of government attempts to enforce the National Wool Grading Standard accepted by the Ministry of Textile Industry, the Ministry of Commerce, and the Ministry of Agriculture, and used by the SMCs in purchasing wool.
- Monitoring quality. For example, each year in the IMAR the FIB takes random samples of wool from SMCs at township, county and prefecture levels (about 20 to 40 units are sampled). Staff from the FIB also observe SMC purchasing practices. At the township level, they check for correct sorting and grading of wool, while at higher levels they monitor regrading and check for possible adulteration. The surveys are conducted in conjunction with the Price Bureau and the Industry and Commerce Bureau and have revealed both poor training among many SMC staff and some deliberate downgrading of wool.

Local or coarse wool is not officially graded in IMAR because, as explained in Chapter 4, the National Wool Grading Standard does not establish any grades for this type of wool. However, in IMAR as elsewhere in China, there are established regional grades or types of coarse wool. Some of the better-known types of coarse wool in IMAR are: Hailaer wool, Chengdu wool, Ingzi wool and Xinin wool. Most local wool is used for carpets and blankets, though Xinin wool with fibres exceeding 20cm in length is used for knitting.

XUAR

The FIB in XUAR is extensively involved in testing and certifying wool. By 1992, about half the wool in the province was being tested and this proportion was expected to increase. Appraisal was both by subjective assessment (for style, colour, fineness, grease and suint, and crimp) and machine testing (for length and clean yield). Tests for fibre strength were only being undertaken at the request of certain traders and were seldom performed. It was claimed that in relation to fibre strength, XUAR wool was more reliable than wool from other parts of China.

In XUAR, the FIB conducts tests on wool being sold by individual State farms, often direct to mills. In some cases, the FIB also tests wool for township-level SMCs. Much of the FIB work, however, is concentrated at the mills, where it supervises the testing of bulk lots of raw wool and replicates the tests on this wool conducted by the mills themselves. The FIB is also concerned with testing wool tops at the mills.

Compared with the TRI outlined in Section 6.6.2, the FIB is more specialised in quality control, the marketing process and some research into industry standards. Thus it has a more administrative role than the TRI which concentrates on research, especially into the chemical and physical properties of various fibres and their products. The FIB also has an authoritative role in settling disputes between mills and wool suppliers, particularly when the dispute goes to court.

Gansu

The Gansu FIB operates under the provincial Quality Control Bureau. More specifically, it is a subdivision of the Wool Product Quality Inspection Agency which is primarily responsible for certifying the quality of wool textile manufacturing products. Founded only recently in 1986, the Gansu FIB has undertaken testing and monitoring of such fibres as wool, cotton, synthetics, hemp, cashmere, camel and rabbit hair, as well as arbitrating on disputes. Compared with the FIBs in neighbouring IMAR and XUAR, the Gansu FIB is newer and its equipment less advanced. The testing of domestically-produced wool is a major part of its activities. Historically, responsibility for wool testing in Gansu has passed from the Gansu No. 1 mill to the Gansu TRI and subsequently to the Gansu FIB.

Despite having the responsibility for and being the logical organisation to perform the testing of wool, the Gansu FIB suffers from being inadequately resourced and the staff lack the necessary skills and equipment. The limited capabilities of the Gansu FIB were highlighted by its "partial" participation in the demonstrations held at the national wool conference in XUAR in 1991 organised by the FIB to promote a pricing system based on clean wool. The Gansu FIB provided samples of Gansu wool but the testing of these samples was carried out by the XUAR FIB. Ultimately, the Gansu FIB is seeking full authority on grading and on the issuing of test certificates.

Although a highly motivated new organisation keen to discharge its duties, it is severely limited in what it can do.

Poorly developed fibre inspection procedures in Gansu Province extend the whole way along the marketing channel, including down to the raw-wool purchase level. For instance, prior to 1990 fine wool Grades I and II were purchased from farmers in most counties as mixed grades. In fact, in Sunan County, which is one of the major fine wool-growing counties in Gansu, wool was purchased on the basis of three broad categories, namely fine wool, improved wool and local (coarse) wool until 1990. This placed the better-quality wool grown in Gansu at a distinct disadvantage relative to wool from other parts of the pastoral region, since elsewhere the wool was at least separated at the time of purchase into the ten grades established in the old National Wool Grading Standard (Section 4.2). The problems traditionally encountered in Gansu with the application of the imprecise old standard augur poorly for any widespread implementation of the new National Wool Grading Standard introduced in December 1993 (Section 4.3).

6.2.3 Fibre Inspection for Imported Wool

As mentioned earlier, imported wools are tested under the authority of the Commodity Inspection Bureau which is part of MOFERT. Tests are primarily carried out for fibre diameter, length, clean yield and vegetable matter, although appraisals for other attributes such as style and colour are now more common. Mills are charged for the inspection, though according to Morris *et al.* (1993) not on a full-cost recovery basis. The Commodity Inspection Bureau is not equipped to carry out all the tests required. Therefore, it issues measurement certificates to the FIBs enabling them to test the imported wool. In effect, the FIBs act as agents for the Commodity Inspection Bureau.

Tests conducted by the Commodity Inspection Bureau or the FIB are in addition to the testing of the wool in the exporting countries. Some arrangements do exist with agencies in exporting countries (such as the Australian Wool Testing Authority and the New Zealand equivalent) regarding the testing of wool to be imported into China. The arrangements enable the testing authority in the exporting country to certify the wool (at the exporter's expense), although random inspections by Chinese authorities may still occur and dispute settlement procedures are in place in China. In relation to Australian wool going to China, Morris *et al.* (1993) cite the cost of this endorsement at AUD$3.60 per bale and a usage rate of around one-third of the wool exported from Australia to China in 1991/92. According to Morris *et al.*, wool-testing procedures in China are not consistent with the international standards set down by the International Wool Textile Organisation. For example, Morris *et al.* mention that the international standards require each bale in a lot to be sampled, whereas the Chinese authorities test only one in every five bales. The non-standard testing procedures followed by the Chinese could lead to incorrect rejection of particular lots of wool.

6.3 Problems in Wool Grading and Fibre Inspection

Much of the concern over incorrect grading of wool has arisen at the raw-wool purchase level. The subjective nature of the appraisal, the broad quality grades and the inflexible price and marketing arrangements have all encouraged abuse of the simple grading system established by the old National Wool Grading Standard. With the fixed grade-price differentials which existed until 1992, SMCs have had an incentive to incorrectly grade in order to avoid a build-up of stocks in "overpriced" grades. In some cases where the county SMCs own and operate the local scouring plant, there has also been an incentive for them to downgrade wool so they could obtain lower-cost wool to supply their early-stage wool-processing operations. Apart from any deliberate manipulation, grading at the grass-roots SMC level suffers from inadequately skilled staff and no equipment, issues taken up in Sections 6.3.1 and 6.3.2. To some extent now that the wool market is formally open to free trade, the competition (or potential competition) from other buyers will restrict the opportunities for the SMCs to exploit their monopoly purchasing power. However, non-SMC buyers of wool are likely to be highly selective both in terms of the areas in which they operate and in terms of the types of wool they are interested in purchasing. The SMCs, therefore, can expect to remain the dominant buyer in most pastoral areas.

Changing the grade as the wool moves through the marketing channel is another problem. For example, Shi (1990) cites the case of the SMC-owned Xinjiang Tea and Livestock Products Company in 1985 which purchased 30kt of wool of which 4.4% was special grade, 30% Grade I and 34% Grade II. However, when selling 22kt of this wool to a textile mill, 17.4% was sold as special grade, 49.7% as Grade I and 20.8% as Grade II. Shi argues impropriety on behalf of the Tea and Livestock Products Company and suggests that this is a widespread problem.

The case to which Shi refers could reflect either a deliberate downgrading of the wool when it was purchased from growers or the favourable outcome of a careful re-sorting and regrading of the previously mixed grades prior to it being sold to the mill. That is, the small parcels of wool which are poorly sorted and, therefore, conservatively graded and priced at the purchasing station may be re-sorted and bulked-up into bigger, more homogeneous and better-graded lines by the township- or county-level SMC. In Australia and in other wool-growing countries, there are specialist wool merchants who perform these bulk-classing tasks for profit. In a free market, there is no reason why the SMCs or other agencies could not find a similar niche in the wool-marketing chain in China.

In most important wool-growing provinces, the FIB conducts surveys to help prevent the exploitation of growers by unscrupulous buyers who deliberately downgrade wool at the initial purchase point, and to encourage a high standard of regrading at higher points in the chain. These FIB surveys are designed to check the grading being undertaken by the SMCs and, after 1992, other units and individuals buying wool. The evidence presented in Section 4.5 suggests that, at least in depressed market conditions such as those in 1991, the FIB has had limited success in controlling deliberate downgrading or incorrect grading by the SMCs.

There is also anecdotal evidence to suggest that the SMCs sometimes incorrectly upgrade wool before offering it to mills because mill managers frequently complained that they found much of the wool they were offered was not as good as the grade it bore would have suggested. For example, in 1992 the manager of the large topmaking mill in Huhehot (the capital of IMAR) claimed that up to 30% of the lots examined by his buyers at the county SMCs were incorrectly graded. The most common problem was that the wool which the SMCs graded as fine wool special grade was not of the length required for that grade and had to be downgraded to fine wool Grade I (or even to fine wool Grade II). It would seem, therefore, that in the past the SMCs have sometimes sought to exploit their monopoly position both by downgrading wool at the time of purchase and by upgrading it when they were selling.

Ideally, all wool ought to be graded according to the Industrial Wool-Sorts Standard prior to scouring. However, scoured wool often does not meet the standard required by topmakers, either owing to technical problems which occur during scouring or owing to incorrect regrading prior to scouring. County scouring operations lack expertise and often use old equipment. Unsatisfactorily scoured wool may be rescoured sometimes, but often is used as it is. If the problem is one of poor grading, the wool has to be blended in with other wool. In a survey of county scouring, the IMAR FIB took eight samples (of 70s count wool) from county scouring plants in August 1990. Five failed to meet the required standards, one because of incorrect wool grading, and four because of poor scouring techniques. No systematic procedures exist, however, to check on the grading and measurements undertaken by classers and technicians at the county scouring plants. The issue of poor scouring and poor grading prior to scouring is elaborated further in Chapter 9.

At the topmaking stage, both the topmaker and the worsted mill grade the tops, at least in the IMAR. If a dispute arises, then the FIB inspects the wool tops. Disputes are generally avoided in normal times since most of the downgrading is done by the topmaker before the tops are manufactured. Topmaking mills in China attempt to overcome the problem of heterogeneous wool through multi-stage blending during the topmaking process. However, good equipment and a high level of skill are required to achieve consistently good-quality tops, given the relatively poor quality of the raw material. Before the "wool wars" (1985 to 1988), around 2 to 3% of tops were incorrectly graded and had to be downgraded by the mill. The percentage of tops which needed to be downgraded escalated to around 30% during the "wool wars". Since 1991, about 5% of tops have been downgraded. Most large worsted mills objectively measure the characteristics of their tops to assist with the control of processing and management.

In disputes between topmakers and the textile mills, it seems the major problem is the weighted average length of fibres. In Lanzhou (the capital of Gansu), for example, the manager of a large specialised worsted mill claimed that there had been a serious decline in the quality of tops in the second half of the 1980s. Specifically, the percentage of Grade I tops (64 spinning count) that had to be downgraded to Grade II or outgrade owing to a weighted average length of less than 72mm, rose from 42% in 1986 to 85% in 1989. Furthermore, 34% of the Grade I tops in 1986 did not meet the limit for the percentage of short fibres, while up to 90% failed this criterion in 1989. Although various reasons for the incorrect grading of the tops were mooted, officials of the mill laid most of the blame on incorrect grading prior to scouring and processing into tops by the small-scale county processors.

Of all the factors instrumental in the incorrect grading and testing of wool in China, a lack of technical skills and testing equipment ranks most highly.

6.3.1 Human Capital

The lack of skills in wool inspection, grading and preparation constrains the implementation of more rigorous quality standards. The rapid development of the Chinese wool industry has outpaced the ability to train the large number of technicians needed to inspect and grade the wool with an appropriate degree of precision. Wu (1990) claimed that more than half the textile mills in China did not employ a single graduate from a formal textile training program.

Some training programs have been available in the past to the traditional SMC buyers and are now to be extended to the new buyers. SMC buyers at the purchase stations are supposed to be checked by the county Animal By-Products Company staff to see they are doing their job properly. The buyers are trained each year and work under a responsibility system. The technician training program is normally organised by the FIB and run jointly by the FIB, MOTI, Ministry of Agriculture and Ministry of Commerce. The duration of the training is usually two or three weeks and it takes place once a year before the wool-purchasing season commences in June. The trainees come from all levels of the SMC and have little or no previous experience in wool grading. The principal task of the training class is to teach the trainees about the National Wool Grading Standard. During the course, trainees are given the opportunity to practise until they are able to grade wool correctly. Trainees receive a certificate on successful completion of the training course. These certificates are supposed to be the minimum qualification required before people are permitted to buy raw wool.

The trainees from provincial-level SMCs are usually taught at central level, and the trainees from prefectural-level SMCs are taught at provincial level. All the trainees from the grass-roots SMCs are trained at prefectural level. Certificates for wool grading are issued by the relevant FIB. That is, the National FIB is responsible for the issue of the certificate to the trainees at provincial level; the provincial FIB is responsible for the issue of the certificate to the trainees at prefectural and county level; and the prefectural-level trainers are responsible for the issue of the certificate to the trainees from the grass-roots level.

The technical college at Wuxi in Jiangsu Province, formally called the "Training Centre for Modern Management of Wool", trains personnel from the textile industry, the animal husbandry production sector and the fibre inspection sector, at both the provincial and prefectural level, in wool inspection, wool grading and wool preparation. It is a department of Wuxi University of Textile Workers and it is partly funded by the Material and Capital Supply Company within MOTI. However, the fees paid by those people attending courses at the technical college are its most important source of finance. About 50 trainees are enrolled in each course and courses run from two weeks to three months. All the instructors are from the Wuxi University. Most courses are held irregularly, depending upon the number of applicants.

The staff of the National FIB are responsible for training FIB personnel at the provincial level in fibre inspection technology. Courses usually last for one month and are held at the provincial level approximately once every two years. Fewer than 50 individuals from provinces all over the country attend each course. Those who attend these courses are expected in turn to train people at the next level down. Subjects taught at each course include animal science, foreign trade commerce, textile science, and biometrics. In addition to central-level FIB staff, teachers from textile universities and colleges are invited to act as instructors at these courses. There were a total of four such training courses held between 1981, when the courses were first introduced, and 1992.

The provincial-level FIB runs training classes for SMC township-level wool graders who can only class wool if they have an FIB certificate. Thus the FIB is involved in testing and issuing certificates to new classers, as well as testing existing classers who can have their certificates taken away if their classing ability does not meet the test. Most county wool technicians have received no formal training although some have been to large mills elsewhere in China for experience. Mill retirees also sometimes go to counties to help, but often they are not the best advisers since their skills are out-of-date.

The kinds of difficulties which can occur because of the lack of suitably trained technicians at the local level are well illustrated by the case of Wushen County of the IMAR which produces some of the better-quality wool in China. In 1989, the Wushen SMC sold wool at the Beijing auction. However, the wool it sold through the auction had to be graded by technicians supplied by the Textile Industry Corporation (TIC) in Huhehot as it did not itself have the capability to prepare the wool according to the auction standard. Altogether, Wushen County sold 500 tonne at the 1989 auction, most of it being fine wool special grade and some being fine wool Grade I. The incentive to participate was the high prices expected at the auction and the publicity the auction provided. The wool offered at the Beijing auction in 1989 sold for ¥19/kg (greasy) compared with ¥11.0/kg (greasy) received for similar wool sold through the traditional SMC channel. Even after allowing for transport costs and the 1% auction selling commissions, the Beijing price was much higher. After 1989, Wushen SMC could not use the auction system because its technicians could not regrade the wool according to the auction standard, and the Huhehot TIC (which collaborated in 1989) was not prepared to supply graders. Instead, in 1991 the TIC bought, regraded and resold at auction about 200 tonne (10%) of the Wushen County clip and made a significant profit on the deal. (See Section 8.3.2 for details.) In 1992, Wushen County SMC officials were planning to send one of their staff away for training and/or to invite qualified technicians from elsewhere to come to regrade their wool so they can participate directly in future auctions. They were also planning to introduce regrading at the township level because this is the logical point in the chain for them to regrade to auction or industrial grades. The main obstacle, however, is a lack of skilled grading technicians.

The need to improve the grading skills in the county became more urgent for Wushen with the establishment of its new wool-scouring plant in 1992. Despite the severe adverse implications of poor grading on the relatively higher-quality wool produced in Wushen County (Section 9.3.2), the need for skilled graders was recognised but not given the priority it warranted. The substantial investment involved in the establishment of scours was complemented only by a consideration to send suitable people away for training, perhaps to the new Training Centre for Modern Management of Wool mentioned earlier, or to invite grading technicians to come to the county to train local graders.

The situation in Wushen is by no means unique and is symptomatic of the situation throughout much of the pastoral region. When scours were established by the SMC in Balinyou County in Chifeng City Prefecture of IMAR, the SMC had no previous experience in wool scouring. More than 100 staff, including electricians, quality inspection staff, production technicians, equipment and management staff, classers and equipment maintenance staff, were sent away for training for periods up to six months. However, expenditure on training in the initial year still amounted to less than 1% of operating costs, and it was uncertain as to what ongoing training, if any, was to be provided.

6.3.2 Testing Equipment

Prior to 1991, there were about 250 wool-testing machines distributed to the FIB offices throughout China. In 1991, another 100 machines were made available. It is estimated that the FIB would need about another 150 such machines (i.e. 500 machines in total) to test all wool produced in China. These machines, which cost about ¥30,000 each, measure clean yield, fibre diameter, and fibre length. The FIB has not sent any machines to the county level, but rather it has concentrated them either at prefectural or provincial offices of the FIB. However, with more and more wool being scoured at the county level, the FIB is grappling with the issue of how to test closer to the grass-roots purchase station.

Perhaps the most important attribute to be measured as close as possible to the point of original purchase is clean yield. Two methods are used to determine the clean-scoured yield of wool in China. The first method, known as the *oven-dry* method, involves: (i) taking the gross weight of a wool sample; (ii) washing the wool sample with water and a degreaser; (iii) drying the wool; and (iv) determining the net weight of the wool sample.

The formula for calculating clean yield is then given as:

$$E_o = \frac{G_s}{G_w} \cdot 100$$

where E_o is the clean-yield percentage;

G_w is the original weight of greasy wool; and

G_s is the weight of clean wool under the rate of moisture regain which is officially accepted (the officially accepted rate of moisture regain is 16% for fine wool all grades and 15% for improved, semi-fine and local wool), and is given by:

$$G_s = G_{wo} \cdot \frac{(100 + W_s)}{100}$$

where G_{wo} is the absolute dried weight of clean wool; and

W_s is the rate of moisture regain for dried clean wool.

The second method, known as the *oil and press* method, involves: (i) selecting a wool sample of known weight (for homogeneous wool, a sample of 200g is used; for heterogeneous wool, a sample of 150g is used); (ii) washing the wool sample of known weight using water and a degreaser; (iii) placing the wool in a press; and (iv) applying a known pressure to the wool to determine the decrease in volume. The change in volume is directly correlated to the clean-scoured yield of the wool sample.

For two reasons, the oil and press method is currently the most widely used method in China for determining clean-scoured yield. First, the time required for the oil and press method is much shorter than that required for the oven-drying method, although it still takes approximately one hour to produce a result. Second, the cost of equipment required for the oil and press method is considerably lower than that required for the oven-drying method. Equipment required for the oil and press method costs around ¥2,000 and is portable.

Technically, therefore, a relatively simple machine costing around ¥2,000 could be made available at the major purchase stations in the main wool-growing areas so that most wool could be bought from growers on a clean-scoured basis. The time taken for each test and the cost of employing a technician to work the equipment both mitigate against the practicality of this approach. However, with modern equipment it is becoming feasible to measure clean yields at the township- and county-SMC levels.

6.4 Quality Control and Fibre Inspection Regulations

Rules governing quality control and fibre inspection are guided by the National Wool Grading Standard outlined in Chapter 4 and by the "Standardisation Law of the People's Republic of China". To gain an appreciation of how these Central government edicts operate in the producing regions, the example of the "Wool Quality Control Regulation for the IMAR" is reviewed in this section. A full translation of this Regulation is provided in Appendix F.

The regulations came into effect in April 1992 at a strategically important point in time, namely immediately prior to the IMAR re-opening its wool market and allowing buyers other than the SMCs to operate. Thus, the timing of the introduction of the regulations, which aim to strengthen fibre inspection activities, appears to have been aimed at avoiding or limiting problems of the kind experienced with wool quality during the "wool wars" (1985 to 1988). The rules also preceded the introduction of a new National Wool Standard and a new Wool Top Standard (Chapter 4).

Nominally, the regulations confer significant additional power on the FIB of IMAR in the areas of wool testing, quality control, inspection, guidance, and arbitration and mediation. Furthermore, the regulations are wide in coverage, with all individuals engaged in wool production, purchasing, marketing and processing being required to comply. Since the FIB is responsible for administering the regulations, the impact of the new rules will be largely determined by the interpretation the FIB places upon them.

The regulations set out strict guidelines for adherence to the National Wool Grading Standard. Grading of wool for purchase outside the Wool Standard guidelines can incur sizeable penalties ranging from substantial personal fines, confiscation of illegal income and wool if adulterated, fines of up to 10% of the gross value of the wool lot, and orders to stop certain marketing activities. These penalties are in addition to any civil compensation claimed by the affected parties or any criminal sanctions. However, the regulations do not specify how adherence to the regulations will be monitored.

The new regulations reflect an attempt to strengthen fibre inspection and to bring more order to the open wool market than was present during the "wool wars". The aim is to avoid a recurrence of the excesses which led to a marked deterioration in wool quality at that time. Whether the standards and fibre inspection agencies are developed enough to achieve this result is debatable. Broadly defined quality standards in conjunction with inadequate fibre inspection capabilities make the regulations largely redundant or at best simply administrative. However, they are an important indication of the desire of governments at the provincial level to accompany broad marketing reforms with stronger institutional arrangements to limit some of the undesirable effects of a free market. As such, they reflect an important step towards recognising the need for a more comprehensive approach to marketing reform.

The IMAR Wool Quality Control Regulation preceded a similar set of regulations at the national level (see Appendix G). This is typical of policy reform in China where important regulations and developments are first tested in the key producing provinces and then modified, or made less specific, to allow for their nationwide implementation. A comparison of Appendices F and G reveals that the National Wool Quality Control Regulation is less constraining and less detailed than the corresponding IMAR regulation on which they are modelled. For instance, the IMAR regulations allow only seven days in which any objections to fibre inspection must be raised compared with 15 days in the national regulations. Furthermore, the procedures and penalties are detailed more precisely in the original IMAR regulations. The subsequent differences observed in the national regulations may reflect evolution of the regulations in response to problems encountered with the IMAR regulations (such as difficulty in implementing the seven-day notification period). However, they are also likely to reflect the need to provide a more general set of guidelines so that the regional authorities have some autonomy to mould the regulations to suit their own fibre inspection, wool-marketing and wool-production systems.

6.5 Selling Wool on a Clean-Wool and Industrial-Grade Basis

To illustrate how the state of fibre inspection in China can hinder some of the agribusiness reforms, consider one elementary reform, though with large potential benefits, namely the sale of wool on a clean-yield basis. The desire to sell wool on a clean-yield basis is clearly reflected in the new National Wool Standard (Section 6.4 in Appendix B) as well as in the recent Wool Quality Control Regulation (Clause V(1) in Appendix F and Section 3 in Appendix G). Indeed, the old National Wool Standard established in the mid-1970s described sales on a clean-yield basis as a "development objective" and as an "urgent need" (Section (II)4 in Appendix A).

Despite the longstanding interest in selling wool on a clean-yield basis, the new standards and quality control regulations which have emerged since 1993 require sales on a clean-yield basis only when large lots of wool are traded. That the new regulations did not extend the requirements of wool sales on a clean-yield basis to the majority of Chinese wool which is traded in small lots reflects, among other things, the current state of fibre inspection facilities in China.

Almost all wool purchased from herders to date has been transacted on a greasy-wool basis, with the only allowance for different clean yields being to shake the wool before determination of the price. Clean yields are particularly low in China, averaging only around 40% in the two main wool-producing provinces of the IMAR and XUAR. Of the more than 1,000 tonne of greasy wool tested in Gansu in 1991, clean yields for the three broad categories were 37 to 42% for fine wool, 40 to 45% for improved wool, and 50 to 65% for coarse wool. The generally low yields are a function of the Chinese wool production systems and especially the environmental conditions in the pastoral region which create a heavy dust load. Management practices are also an issue, as neighbouring households achieve significantly different yields. For instance, clean yields on some of the better State farms and stud farms exceed 60%. Thus, it would seem that provided the right incentives are in place, clean yields and hence wool quality could be improved significantly. The current system of pricing on a greasy-wool basis provides no such incentive; indeed it is a disincentive.

The FIB is conducting extensive research and development with a view to introducing a wool-pricing system based on clean-scoured yields. However, the purchase and sale of wool on a clean-scoured-yield basis is still at an experimental stage. In 1991, as mentioned in Section 6.2.2, the FIB held a national wool conference in XUAR to promote a pricing system based on clean wool. In the medium term, some officials claim that the FIB hopes to increase the amount of wool sold on a clean-yield basis from the current level of around 10% at the start of the 1990s to around 70% in the forthcoming five years. Estimates and expectations vary, however, depending on the officials contacted; for example, Zhang (1990a) cites a figure of around 30% by 1995. Individual villages will be encouraged to produce greater quantities of homogeneous wool. A single clean-yield test can then be conducted on a bulk lot of wool from a large number of households. Such an approach could reduce the requirement for large numbers of testing machines and/or large lead times between the time of delivery and when test results are returned to individual households. It is anticipated that the broad-based introduction of a clean-wool pricing system will take place first in the extensive pastoral areas and then feed into the less extensive semi-agricultural areas such as Aohan County in Chifeng City Prefecture.

The sale of wool on a clean-wool basis is strongly supported both by the former MOTI and by MOA. Lower clean yields, wrongly graded lots, and increasingly shorter fibre lengths are a few of the problems with domestic wool faced by the textile industry in recent times. The main reason why MOTI has supported the clean-yield experiments is that it would address one of these quality problems. However, more fundamental benefits may arise in other ways. Sale of wool on a clean-scoured basis would facilitate the introduction of alternative and more direct marketing channels. The MOA support for the experimental pricing system may also reflect a desire to avoid the inefficiencies of the SMCs as well as to improve, in the longer term, wool quality and incomes of herders. Nonetheless, support for the notion of clean-wool selling cannot be generalised too much and depends on the relative power and innovativeness of the institutions involved. For instance, in the XUAR, the SMC joined other government agencies in advocating that all wool be purchased on a clean-scoured basis from 1992.

To date, there have been a number of trials conducted on the feasibility of the purchase and sale of wool on a clean-scoured basis. The trials have predominantly focused on the use of two different marketing channels, namely direct sale between producer and processor, and indirect sale by use of the auction system. With regard to direct sales, processors are being encouraged to purchase wool directly from either State farms or from individual villages in production areas deemed to have sufficiently large amounts of homogeneous wool available. Currently, only villages which produce more than 500kg per annum of wool of the same improved level are being considered for direct-sale trials. Villages with lower production levels are not eligible to participate in the trials, largely because production levels below this amount lead to an unacceptably high level of heterogeneous wool, thereby increasing the number of clean-yield tests required for an objective assessment of the wool offered for sale. Villages participating in the trials are located in the major wool-producing provinces of IMAR and XUAR, Gansu, Jilin, and the southern Chinese province of Zhejiang.

In Baicheng Prefecture of Jilin Province, the entire wool clip is sold directly to wool processors on a clean-yield basis. In this prefecture, as in other areas selling directly to processors on a clean-yield basis, herders are paid on an estimated clean yield. The wool is then tested by the prefectural FIB and any difference between the prices based on the estimated and actual clean yield is paid to the wool producers. The clean yield of the wool is determined from a sample taken at the purchase station by the prefectural FIB inspector. There are two methods used to obtain the sample, one being the open-package method and the other being to punch a hole in the bag in a random spot as is done in Australia. The price of wool purchased on a clean-yield basis is determined by the existing price structure for greasy wool, assuming a clean yield of 40%. This assumption provides the reference point for increases or decreases in price based on the actual clean-yield results. Each percentage deviation in clean yield from 40% attracts a change of 2.5% in the price for the relevant grade of wool on offer.

Direct sales by farmers to processors based on clean-wool yields have only been conducted in fine wool production areas. Hence, the 40% clean-yield assumption is appropriate because the clean yield of fine wool grown in China averages about 40%. It would not be appropriate to use 40% as the reference yield if the wool being purchased was semi-fine since this type has an average clean yield of between 45% and 55%.

Wool sold by State farms directly to the mills can also now be sold on a clean-wool basis. The FIB does the testing of this wool and the price paid depends on the clean yield.

The volume of direct sales varies from province to province. In the XUAR the proportion of the total clip sold directly to mills is less than 25% and most direct sales are by State farms. The August 1st Mill in Shihezi City in XUAR (which is under the control of the PCC) has purchased wool on a clean-yield basis since 1988 both from PCC farms and from the SMC. Wool is tested by the prefectural FIB under the Standards and Management Branch of the Shihezi prefectural-level government. Final payments for the wool are made after testing. The nearby Xinjiang mill at Changji is not keen to price its purchases on a clean-wool basis as it claims the equipment which measures clean yield is not sufficiently accurate with a variation of 3 to 5 percentage points, and the mill uses it only as a reference rather than a basis for the final price.

Implementation of a more general and widespread system of clean-wool selling through the SMCs would be difficult because most township and county SMCs have no experience with buying wool on a clean-scoured basis. The system in use in Baicheng Prefecture of Jilin Province could become a model for other areas. Under this system, farmers have their wool bulked up after grading at the purchasing station. Each lot of wool of a particular grade from each purchasing station is tested for clean yield by the FIB at the prefectural level. Farmers with wool in that lot are paid on a clean basis pro rata. That is, the purchasing stations are acting as wool-pooling agents. It was claimed that the FIBs in the main wool-growing prefectures in XUAR all had the equipment to measure clean-scoured yield and that the Baicheng system could be applied without too much difficulty.

Farmers in Wushen County in IMAR are not paid on a clean-scoured basis. However, some of their wool is subsequently sold on this basis. Apart from the wool sold at auction in 1989 (as discussed in Section 6.3.1), Wushen County SMC sold 50 tonne to a Nanjing topmaking mill and 100 tonne to a mill in Shenyang on a clean-scoured basis in 1992. The wool was scoured at the Nanjing and Shenyang mills before the final price was determined. Subsequently, Wushen County has not been involved in other sales on a clean-scoured basis. However, it does intend to sell on this basis in the future. If the full benefits of clean-wool pricing are to be realised, however, then it must be extended to purchases from individual herders as the ultimate supplier of cleaner or better-quality wool. However, Wushen County SMC has no equipment to enable it to buy from the farmers on a clean-scoured basis. It is planning to introduce regrading at the township level in future, which is a logical point in the chain for it to regrade to auction or industrial grades. Indeed, three townships in Wushen County have been involved in experimenting with clean-wool sales between the township SMC and a number of mills in the provincial capital, Huhehot. However, the main problem of a lack of skilled grading technicians remains.

Even in some semi-agricultural areas such as Aohan County in Chifeng City Prefecture in IMAR, experiments are being conducted on clean-wool pricing. The Aohan County SMC, in conjunction with the prefectural FIB in Chifeng City, is presently experimenting with the sale of wool on a clean-scoured basis to the No. 1 and No. 2 textile mills in Chifeng City. The experiment presently being conducted is only the first stage in a much more ambitious longer-term plan whereby households will be able to sell their wool to the SMCs on a clean-scoured basis. For logistical and financial reasons, it was decided to experiment first with the easiest part of the wool-marketing chain, that is, the sale of wool between the county SMC and the processor. The experiment essentially involves setting prices according to a standardised clean yield as outlined previously and then making a retrospective adjustment once the actual clean yield has been determined. The main difference is that the standardised clean yield adopted for Aohan County is said to be 34 to 36% compared with the base clean yield of 40% recommended by the National FIB and used in other parts of China. Wool with these particular clean yields attracts 100% of the normal purchase prices paid. For each percentage point below 34% clean yield, the price paid for the greasy wool is decreased by 2.5%. Conversely, for each percentage point above 36%, the price is increased by 2.5%. The wool is tested at the county-level SMC by staff of the Chifeng City prefectural FIB.

There have also been some experiments with selling on a clean-yield basis in Gansu. The Gansu Textile Industry Buying and Marketing Bureau, together with the Gansu Quality Management Bureau (Fibre Inspection Division), the Gansu Price Bureau, the provincial-level SMC, and the Light Industry and Textile Industry Department of the provincial government, all agreed to conduct a series of experiments in two counties in 1990. These counties were Huang Cheng County, which is in the area east of the Yellow River and which grows mainly coarse wool, and Sandan County which produces both fine wool and coarse wool. (Huang Cheng is also the name of the top fine wool stud in Gansu but this stud is located in Sunan County well to the north west of Huang Cheng County. A full description of Huang Cheng stud may be found in Longworth and Williamson, 1995.)

For the Gansu experiment, testing equipment to measure clean yields (which was said to cost about ¥2,000 per set) was to be located in the offices of the county-level SMC. If these experiments proved successful, the Gansu authorities had plans to expand the objective measurement of wool at the county-SMC level to all counties producing more than 50 tonne of fine wool in the near future.

While measuring the clean yield before setting the price for a wool lot is a major step forward, it would be even more useful if the wool was properly graded prior to the measurement of the yield. In this regard, it is interesting that some of the major State sheep studs, such as Gongnaisi in XUAR, are already selling their clips directly to processors not only on a clean-scoured basis but also on the basis of the Industrial Wool-Sorts Standard. Indeed, in 1992 it was claimed that about one dozen State farms were selling their wool direct to mills after grading it according to industrial grades.

Although it is impracticable to expect that wool could be graded according to the Industrial Wool-Sorts Standard at most purchasing stations, the importance of good grading prior to scouring suggests that wool needs to be graded into industrial types (or at least into grades which are much closer to the requirements of the processors) before the clean yield is measured and before the purchase price is finally determined. This could be done at the township-SMC level in major wool-growing areas and at the county-SMC level in other areas. That is, the native goods companies which handle wool for the SMC at the county level and above need to become bulk-handling and re-sorting agencies similar to the private firms which undertake this role for wool growers with small clips in Australia. It seems the Wushen County experiments in the IMAR and the Huang Cheng and Sandan County experiments in Gansu could be a major step in this direction.

6.6 Fibre Research

Fibre research and development have been integrally tied to fibre inspection in China. Not only have fibre research institutions been involved in developing new products for the Chinese wool textile industry but they have also been directly involved in fibre inspection issues. They have played a role in the development of the various industry standards and they have also undertaken fibre testing. Through a brief overview of some key fibre research institutes in both the wool-producing and wool-processing regions, this section outlines the link between fibre research and fibre inspection.

6.6.1 Fibre Research at the National Level

The Beijing Wool Textile Research Institute (BWTRI) was established in 1979 and is the "key" wool textile institute in China. It is under the Beijing Municipal Committee and has close links with MOTI. A similar large wool research institute exists in Shanghai and a smaller one operates in Tianjin.

There are three main centres under the BWTRI, namely: China National Wool Fabric Test Centre which is an International Wool Secretariat-accredited test centre supervising the testing of quality of wool fabrics for the whole country; China National Wool Product and Investigating Centre which develops new products; and China National Wool Information Centre which issues the *Wool Textile Journal*. MOTI commissioned BWTRI to run these three centres and provides funding for them.

The Beijing Research Institute also has a research and development centre with a pilot plant used as the research base of the Institute. The Institute has been involved with major external aid and development projects such as the "China Wool Textile Technology Transfer Project" funded by the Australian International Development Assistance Bureau (AIDAB) which investigated wool specification, vegetable matter removal, and dyeing and finishing. In general, the Institute carries out many research projects examining the applicability of international technology to meet Chinese needs. Current and proposed projects include new applications and developments in relation to special animal fibres (cashmere, yak, rabbit, angora), modified wool, cotton/wool blending, wool/jute blending and wool/chemical-fibre blending. Research is also conducted into machine-knitted yarn, shrink-proofing and new technology for spinning. Much of the Institute's work is based on the perception of a greater demand for fine wool in the future, as wool products and wool/cotton blends become lighter and require finer wools.

Prior to 1980, research at the Institute concentrated on the chemical and physical properties of domestic wool. However, during the 1980s the Institute became more involved with the testing of domestic and imported wool. The BWTRI is closely aligned with MOTI and acts as one of the two final arbitrators in disputes over imported wool. In this regard, there is room for confusion over the respective roles of the FIB and the BWTRI. If the FIB finds any problems after it tests a consignment of imported wool, the Textile Industry Corporations or other buyers of the wool will take the problems up with the importing agency, such as CHINATEX (China Textile Import and Export Corporation). However, if the wool is cleared through to the mill and the mill subsequently discovers a problem, then the wool is sent to BWTRI for testing. In essence, the FIB deals with the purchasers of the imported wool, whereas the BWTRI deals with the ultimate end-users, that is the mills.

6.6.2 Fibre Research in the Wool-Growing Provinces

XUAR

The XUAR Textile Research Institute (TRI) is involved with research into both fabrics and raw materials. Specific areas of research it has engaged in include physical and chemical properties of fibres; development of new products; and sheep markers. (The Institute has a small plant for producing dye used to mark sheep and that can easily be scoured out of the wool. At present, many wool growers persist in using tar to mark their sheep. Tar [or asphalt] is extremely difficult to remove by scouring.) Because the XUAR produces few synthetics, the Institute conducts little research in the area of wool blending with synthetics.

The TRI is not involved in testing wool which is being traded directly between producers and processors but has been involved in testing wool destined for auctions. Furthermore, it tests wool from selected breeding farms and, in particular, wool from the two national studs in XUAR, namely Ziniquan and Gongnaisi (Longworth and Williamson, 1995). These tests are conducted to assist with the breeding programs at the studs. In the past, the XUAR TRI had a centre for testing the physical and chemical properties (dyeing, strength, density, shrink ratio) of all types of fabrics (wool, cotton, etc.). However, these tests are now conducted by an independent organisation under MOTI called the XUAR Quality Testing Centre for Textile Products. The TRI in XUAR exchanges information with the BWTRI through the *Wool Textile Journal* and also has sent staff to Beijing for training.

Gansu

The Gansu TRI has pilot production lines for wool knitwear, synthetics and worsted fabrics. The pilot plant can produce 150,000m of worsted fabric, 200,000m of woollen fabric and 200 tonne of synthetics per year. For the pilot woollen mill, the TRI starts with scoured wool and for the worsted line it starts with tops. There is a total of 600 people on the staff of the Gansu TRI, with 80 of these being researchers. The Institute has also been engaged in testing and monitoring for the provincial Textile Industry Corporation. The Gansu TRI interacts with the BWTRI with respect to testing technology and also to obtain fashion designs.

In contrast to the TRI in XUAR, the Institute in Gansu does experiment with synthetic fibres, especially polyacrylonite. In respect to blends, researchers at the Institute were experimenting with 60s- and 64s-count wool in a 55:45 wool:synthetic blend for worsted products. For woollen blends, improved fine wool Grades I and II were being blended in a 60:40 (wool:synthetic) or 50:50 ratio. Blending was undertaken both to increase the average length of the fibres and to reduce the cost. The increased emphasis on blends in Gansu reflects the greater production and use of synthetic fibres in that province. As at the other TRIs, little or no research was being conducted in the early 1990s on why length, clean yields, percentage of short fibres and processing conversion factors were all declining for domestically-grown wool. This is surprising given the emphasis mill managers placed on these problems.

IMAR

Research work at the TRI in IMAR is at three levels, namely: raw wool to garment; basic research on wool fibres; and research on other animal fibres, including camel cashmere. The Institute generates some of its own research funds, but also obtains some government funding for its research proposals. The Institute has its own pilot mills for research purposes, but also conducts joint research with commercial mills.

One of the problems for wool quality in the IMAR has been the rapid development of county scouring facilities (Chapter 9). Some of the difficulties facing county scours derive from their use of local chemicals rather than more expensive but much better-quality imported detergents. The local detergents are more difficult to control, especially given the relatively low management ability of technicians at county scours, and often lead to tangled scoured wool owing to the use of wrong temperatures in scouring and inappropriate use of chemicals. Chinese wool grease is often oxidised and therefore hard to scour out. Both the high dirt levels and the poor skirting practices in China act to lessen the effectiveness of scouring. Wind, sun and scouring temperature all act to cause yellow grease. For domestic wools, a scouring temperature of 48°C to 50°C should be used, compared with 48°C to 52°C for Australian wool. Higher temperatures cause excessive cotting of domestic wool, with the percentage cotted increasing from around 2 or 3% to 15% cotted wool over a range of only 1°C at the critical threshold level. The lower temperatures which must be used for domestic wools mean that scouring is less effective. Although cotted wool can be combed out, the top will still have many short fibres. In most cases, badly cotted wool can only be used for woollen products, and usually for felted overcoats. The TRI has done some research on detergent use but does not monitor scoured-wool quality.

6.7 Concluding Remarks

Developments in wool fibre inspection and testing do not appear to have kept pace with the other rapid developments in the commercialisation and modernisation of the Chinese wool industry. Given the limited resources in many regions, there is a need to coordinate better the existing fibre inspection and testing activities as there is considerable overlap. Part of the problem arises from the different official wool standards which are applied at different points in the marketing chain and which encourage different and duplicate testing.

While opportunities for alternative trading channels opened up with the reforms of the early 1990s, there is a need to develop innovative approaches to fibre testing and grading to facilitate these new channels. With the rapid developments occurring in the Chinese wool industry, however, there is a real risk of developing *ad hoc*, overlapping and inaccurate fibre inspection systems. Given the paucity of development funds likely to be directed at fibre inspection and testing, there is a pressing need to rationalise and coordinate activities in this area.

The recent deregulation of wool markets may exert a mixed impact on fibre inspection, testing and grading. On the one hand, a pricing system which allows more appropriate grade-price differentials to develop may encourage further refinement of the current industry standards and stimulate the investment in fibre inspection and testing which would be needed to support a more sophisticated raw-wool grading system. Conversely, free markets with a larger number of relatively inexperienced buyers may lead to even more difficulties in checking and monitoring grading, as evidenced in the extreme during the "wool wars".

Chapter 7

An Agribusiness Giant Awakes

Appointed by the State as the official purchasing authority for raw wool under the past regulated arrangements, the Supply and Marketing Cooperatives (SMCs) traditionally occupied a prime position among agricultural "cooperatives" in China. The SMCs are a true agribusiness giant operating over 63,000 agricultural-product purchasing and procurement agencies throughout China at the start of the 1990s as well as around 10,000 warehouse and transport agencies. As shown in Fig. 7.1, they have handled up to 176kt of raw wool per annum. In many ways, however, the SMC organisation has been a sleeping giant. Cosseted by monopoly purchasing powers, it has had little incentive to adapt to changing market conditions or to improve the way it operates. Indeed, often it has been more concerned with making cosy arrangements for itself and its political masters than with conducting its commercial business.

The agribusiness reforms have brought a relatively sudden and rude awakening for this sleeping giant. The wool-marketing reforms of the mid-1980s, the subsequent chaos of the "wool wars", the development of wool auctions and experiments with direct sales between State farms and mills have all eroded the absolute volume of wool business handled by the SMC (Fig. 7.1). The China-wide opening of wool markets in 1992 has further dramatically reduced the SMC share of the wool market to around 50%, and its purchases of 72kt in 1993 were over 100kt less than a decade previously. The question then arises as to whether this sleeping giant is about to become a dinosaur—a well-known relic of a past era—or whether it is to return to its pre-eminence as a marketing cooperative with a dominant role to play in the marketing of wool. Various factors suggest that the SMCs can survive and indeed can form an integral part of the post-1992 wool market in China. If they are to survive, however, they will need to modernise and to change their practices. Outside (foreign) entrepreneurs could help with this modernisation to mutual advantage.

Initially, the chapter describes the sleeping SMC giant: what it is, what it does and how it acts. The alternative marketing channels which emerged during the 1980s and which have placed enormous pressure on the wool business of the SMCs are then described. How the SMCs have reacted to these agribusiness reforms and, in particular, to the deregulation of wool markets in 1992 is also discussed.

7.1 SMC Status

When the Central government designated wool as a category II commodity in 1956 (Section 3.1), it appointed the network of SMCs as the official State procurement agency. That is, the SMCs were responsible for purchasing and collecting the State production quota on behalf of the State and, in most parts of China, for purchasing all over-quota wool. Originally, the SMCs were established as genuine cooperatives in

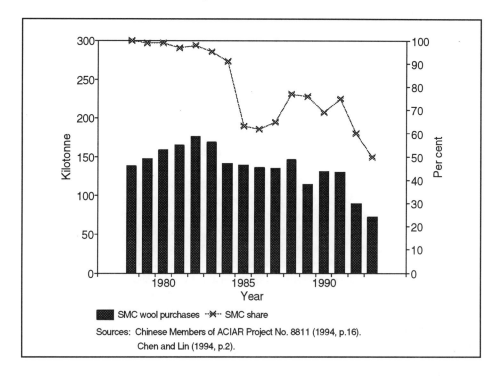

Fig. 7.1 SMC Wool Purchases and Share of Total Wool-Buying Business, 1978 to 1993

the first half of the 1950s. As these cooperatives merged at county, prefectural or city, provincial and national levels, they became closer to the State. After 1956 and especially during the 1960s, the SMC network was absorbed into the State bureaucracy, initially in parallel with the vertical administrative structures responsible to the Ministry of Commerce but eventually becoming part of this vertical hierarchy. During the 1980s, there was a renewed interest in having an independent, farmer-controlled, agribusiness cooperative movement in China (An, 1989). Some provinces, such as Gansu, encouraged the SMC to break away from the State bureaucracy and become more independent. Nevertheless, in most provinces, and even in Gansu, the SMC and its associated companies such as the animal by-products or native goods companies (which trade in wool on behalf of the SMC) remain an arm of the State under the control of the Ministry of Commerce or Ministry of Internal Trade as it has been called since April 1993.

Thus the SMC appears to be a kind of quasi-government organisation. It is neither an independent cooperative in the true sense of the word, nor is it a conventional State enterprise. In principle, the SMC is a cooperative at the township level. That is, individual farm households have at some time in the past joined a township-level cooperative and have paid their membership subscriptions to this cooperative so that they became shareholders at this level. Currently, however, the township-level SMC is not a true cooperative. Instead it is a quasi-government organisation vertically controlled by the Ministry of Internal Trade. Horizontally, it owes political and bureaucratic allegiance to the township government; at the next level up the allegiance is to the county government; at the next level up to the prefectural government; at the next level up to the provincial government; and finally it is part of a Central government ministry. At some time in the past, the farmers who were then genuine shareholders in a cooperative at the township level had the right to elect freely the officials of the SMC at this level. It appears that in many areas the farm households have retained the right to elect the officials who run the township SMCs. However, the elected officials must have the approval of the township government. At the county level, the officials are again nominally elected by a kind of congress which consists of people elected from the various township SMCs. Once again these officials are essentially appointed by the county government, since they must have official approval.

Some examples will provide insight into the diverse nature of the SMCs, especially at the lower levels. In Aohan County in Chifeng City Prefecture of IMAR, the SMC operates essentially as a true cooperative movement, with about five-sixths of all households in the county having shares in township cooperatives. Households are permitted to join and leave the SMCs whenever the time arises. In addition, households choosing to leave the cooperative are permitted to withdraw any equity that they may have in the cooperative. Equity in the form of shares is said to provide two kinds of financial rewards: a return equivalent to the long-term bank interest rate; and an unspecified share of the business profit from the branch-level SMCs. The SMC must earn a certain level of profit (a kind of profit quota) before it is permitted to pay dividends to shareholders. That is, dividends are payable from over-quota profit only. The minimum shareholding in the SMC is that number of shares equivalent to ¥2. Workers employed in the SMC are required to have a minimum shareholding in the cooperative of ¥100.

As outlined earlier, the management of each level of the SMC is in theory elected and directed by the grass-roots base of the cooperative organisation. The county-level SMC in Aohan County is administered by a board of directors known as the Cooperative Affairs Management Committee. Two-thirds of the 41 members on this Committee are farmers, with the remaining one-third being elected government officials and SMC managers from the branch-level cooperatives. Committee members are elected by a general assembly of grass-roots cooperative members for a term of three years. The general assembly meets every year to consider, among other things, new appointments to the Cooperative Affairs Management Committee. Candidates for election to the Committee are first nominated by a Presidium elected by the general assembly for the SMC. Members of the Presidium are elected annually at each general assembly meeting. The County government and the County CCP play a major role in nominating members of the Presidium, as it is this body which is the key to the election of the most "appropriate" members to the Cooperative Affairs Management Committee.

The SMC in Cabucaer County in Yili Prefecture of XUAR provides another illustration and demonstrates the diversity which exists amongst county-level SMCs. In Cabucaer, the SMC is controlled by the Farmers' Representative Council or Assembly which elects a Board of Directors for the County SMC and a Monitoring Committee to check on the activities of the Board of Directors. The staff of the SMCs at county and township levels are not government employees. Until 1983, their salaries were paid from the county-government budget but since then the SMCs have been independent. In Cabucaer County, the SMC has resumed its original form of being a genuine farmers' cooperative working closely with the county government but as an independent organisation. Although this is National government policy, it has been implemented to different degrees in different areas.

The SMC staff in Cabucaer are paid from an SMC staff salary fund which is financed by the profits generated by two SMC companies. The various levels in the SMC work under a contract with the next level up in the system. That is, the county SMC signs a contract with the township SMCs and the two companies. The purchase stations and some branch shops sign contracts with the township SMCs. At each level, targets are spelt out in the contracts. If all targets are achieved, then staff get 100% of salary. If targets are exceeded, they can earn sizeable bonuses. If targets are not achieved, salaries can be reduced.

7.2 SMC Structure and Role

The hierarchical structure of SMCs is indicated in Fig. 7.2. The grass-roots or basic SMCs are organised at the township level. Each township SMC is a member of the county-level SMC or a so-called third-level SMC. In turn, the county SMCs are members of the prefectural SMC which is referred to as a second-level SMC. The prefectural SMCs belong to the provincial-level SMC which should logically be referred to as a first-level SMC but this term is not used. At the central level, the provincial SMCs form the All China Union of SMCs.

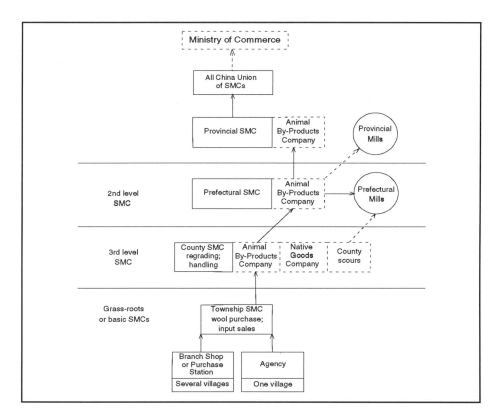

Fig. 7.2 Structure of the SMC and Traditional Flows of Wool

7.2.1 Township SMC and the Scale of Wool Purchasing

A township SMC will have a branch shop and/or purchasing station or depot to service most administrative villages under the township government. Since each administrative village may consist of several physically separate natural villages, the SMC may also establish agencies in these natural villages. The SMC in Aohan County, which was one of the examples cited in the previous section, has 25 basic SMCs at the township level which altogether operate 84 branch shops and 40 agent shops. Most of the purchasing of wool is done by the branch shops, as agent shops usually do not purchase wool. However, the basic SMCs in the townships, the branch shops and the agent shops all sell farm inputs and other household goods.

Besides acting as the traditional State purchasing authority for certain farm outputs such as wool, the SMCs also operate like general stores, selling processed food (powdered milk, biscuits, cakes, etc.), alcohol, clothing, household tools and consumer durables (radios, bicycles, fans, TVs, etc.) as well as certain farm inputs such as fertilisers and pesticides. Private shops are also permitted to sell items for daily living but until the 1990s these shops could not compete with the SMCs in the sale of production inputs.

Traditionally, the SMCs have not been permitted to sell certain farm inputs such as most seeds (there are special seed companies which still retain the State monopoly for the sale of most seeds) or petrol and oils (the State fuel companies still retain this right). The SMCs were also not permitted to purchase grains or meats from farmers, since there were special State monopolies established under the old Ministry of Commerce to purchase these commodities. Generally, however, the SMCs were allowed to purchase most agricultural commodities from farmers once the farmers fulfilled their official production quotas. Nowadays, with freer markets, the SMCs have become much more involved in trading some of these commodities.

The number of SMC purchasing stations in any county will depend on a number of factors, including the number of administrative units in the county, the amount of wool (and other products) to be purchased and the location and dispersion of production. For example, in Wushen County in IMAR, which is a major fine wool-growing pastoral county, there are 14 township or grass-roots SMCs and these grass-roots SMCs operate 87 wool-purchasing stations. The largest purchasing station buys around 80 tonne and the smallest about 20 tonne of raw wool each year. The most important wool-buying township SMC handles around 400 tonne of raw wool each year, while the least important handles about 100 tonne. By contrast, in Aohan County which is also a major fine wool-growing county but a county in which wool is not nearly as important as it is in Wushen, the largest amount of wool purchased at any of the purchasing stations in any one year is about 75 tonne and the smallest amount is only 3.5 tonne.

Cabucaer County in XUAR is an even more agricultural and less pastoral county than Aohan. Yet in this county the wool-purchasing stations or depots are larger and more specialised. Cabucaer County produces around 500–700 tonne of greasy wool each year, excluding wool produced on the Cabucaer State Farm. It is difficult, however, not only in Cabucaer but in most counties, to predict how much the SMC purchasing stations will buy, because farmers living near the county border and close to a purchasing station in a neighbouring county may prefer to sell their wool through that station. Similarly, especially in the case of Cabucaer, some farmers who live in neighbouring counties may elect to sell their wool to the Cabucaer County SMC purchasing stations. Thus quantity produced and quantity purchased in a county is unlikely to be the same even when the SMC is supposed to have a complete monopoly over the buying of wool. Eight township SMCs operate under the Cabucaer County SMC. Within the township SMCs, there are six animal-product purchasing stations, the largest purchasing station buying around 100 tonne of raw wool per year, the smallest about 30 tonne per year.

Apart from the need to cater for a much smaller total wool production in Cabucaer County than in counties such as Wushen, the fewer and larger purchasing stations in Cabucaer also reflect an experiment with centralised wool selling. In recent years, all wool has been purchased on the basis of the old National Wool Grading Standard. Most farmers bring their sheep to be shorn to one of the shearing sheds attached to the six purchase stations. Each of the purchase stations operates two or three shearing sheds in which the sheep are mechanically shorn by professional shearers. The wool is then graded by an SMC technician and checked by an AHB technician. That is, the farmers' wool is carefully graded at shearing time. In 1992, the shearers were charging ¥0.60 to shear each sheep and there was also a further ¥0.40 per sheep charged for electricity. The shearer's charge may rise or fall with the price of wool. The SMC at the township level pays some of this shearing cost (perhaps 70%) to encourage farmers to use this system for shearing. Shearing the sheep professionally on clean boards improves the quality of the fleece and reduces contamination of the wool. The centralisation of shearing also makes grading more effective and efficient. The introduction of centralised shearing and grading in Cabucaer highlights how the SMCs have played an active role in the modernisation of Chinese wool marketing at the grass-roots level.

Sunan County, which is one of the major fine wool-growing counties in Gansu Province, also illustrates how the county SMC has concentrated wool purchasing at relatively few points. However, in this case, the SMC has been remarkably slow to adjust to a more sophisticated grading system. Until 1990, it was content to purchase the best fine wool grown in Gansu on a mixed-grade basis. There are six district (township) SMCs in Sunan which together purchase about 1,500 tonne of raw wool each year through eight wool-purchasing stations. The largest purchasing station buys around 450 tonne while the smallest would purchase around 50 tonne. By way of contrast, in Dunhuang County, which is not a major wool-growing county in Gansu, the county SMC operates 27 purchase stations but only one or two of these buys a significant amount of wool.

7.2.2 County SMCs and Wool

The township or grass-roots SMCs purchase the wool from the growers at the purchasing stations as described in Section 6.1, and then transport it to the SMC at the county level. As mentioned earlier, the county SMCs not only have allegiance to higher- and lower-level SMCs but also have close links with the county government, and many of their activities and decisions reflect these links.

County-level SMCs usually have associated with them specialist companies to handle certain specific products or tasks. These companies form part of a vertical network with corresponding specialist companies at the prefectural, provincial and national levels. For example, wool is primarily handled at the county level and above by animal by-products companies (or native goods companies in minority-nationality areas). Other examples of SMC companies include fruit companies and the tea companies. The county Animal By-Products Company is responsible for selling the wool to other agencies and units on behalf of the SMC. These specialist companies usually perform like State-run organisations, although, as with their parent SMC organisation, they are not unambiguously State instrumentalities. The animal by-products or native goods companies handle wool, hair, hides or skins but they do not handle meat, milk, fruit, vegetables or grains. They can be seen as a division of the SMC organisation which physically handles certain animal by-products. Specialised marketing services are usually required at the county level and above and hence the animal by-products or native goods companies begin at the county level. However, if there is little demand for these services at the county level, the animal by-products or native goods companies may operate instead only at the prefectural or city level. These companies are a significant contributor to SMC activities and accounted for 10 to 12% of the total purchase value of all products acquired by the SMC in China in 1989.

The SMCs at the township level act as agents for the specialist companies. For example, in the case of wool, the township SMCs purchase the wool from growers through their purchasing stations and then deliver it to the animal by-products company at the county level, which in turn delivers the wool to the prefectural Animal By-Products Company. In the past, the prefectural company would sell the wool to processors with a fixed mark-up on the price paid to growers, and the township-level SMC and county-level SMC would each receive a specific percentage commission for handling the wool (Section 5.2.1). With the opening of the market to other buyers since 1992, the fixed-percentage mark-up pricing arrangements of the past have come under great pressure.

The SMC in Aohan County, which has already been referred to several times in this chapter, is a typical county SMC. It has four specialist companies associated with it, namely: the Side-Foods (condiments, sauces and flavourings) Company; the Animal By-Products Company; the Agricultural Input Goods Company; and the Trade Centre Company. The county SMC is headquartered in the county capital, Xin Hui, and as mentioned earlier, it has 25 basic SMCs at the township level, 84 branch shops and 40 agent shops. The SMC in Sunan County in Gansu Province, on the other hand, is less typical since it has only one company associated with it, namely the Comprehensive Purchasing and Marketing Company. Apart from purchasing wool, cashmere, hides and other agricultural products in which the SMC is permitted to trade, the company is also concerned with wholesaling industrial goods to other outlets in the county and distributing, as a wholesaler, alcohol, tea, fertiliser, household goods and tools, electrical goods and spare parts.

Apart from handling raw wool, county SMCs have become involved in other activities related to wool. Many of these activities are closely connected with the interests of the county-level governments. Fiscal reforms in China in the early 1980s made county governments more responsible for their own revenue generation. In many of the pastoral counties, there are few feasible revenue-generating activities other than the processing of minerals and animal husbandry products such as wool. The fiscal reforms of the early 1980s, therefore, renewed interest in the development of local wool processing, not only in early-stage processing plants (wool scouring and topmaking) but also in small woollen mills and garment-making factories.

In Chifeng City Prefecture, for instance, seven out of the 12 counties have factories producing knitted goods, while six out of the 12 are involved with scouring wool. (See Chapter 9 for more details.) In addition, there are a number of leather factories run by the county SMCs as well as local enterprises handling leather. Aohan County SMC, for example, apart from its wool scours, operates a "side-foods" processing factory that manufactures food flavourings as well as a knitting factory which produces synthetic knitted cloth for making wool packs and other bags. In neighbouring Balinyou County, an administrative unit under the SMC called the Industry and Trade Corporation has a number of animal-product-processing companies under its control including a wool scour, pullover factory, cashmere combing factory, leather-processing factory, and a garments-making factory. Altogether, the Balinyou SMC employs around 2,000 people, 700 of whom work for the Industry and Trade Corporation or its associated factories.

The extent of wool-related activities varies across counties depending on the importance of their wool production and other activities. In Cabucaer County in XUAR, the SMC operates two companies, namely: the SMC Trading Company; and the Tea and Animal By-Products Supply and Marketing Company. The county does not have any SMC-controlled animal-product-processing plants, nor is it involved in regrading wool because, as explained earlier, the wool is properly graded at the centralised shearing and purchasing stations. However, it does operate a tobacco-processing plant.

Because the SMC traditionally had a complete monopoly over wool supplies, most county-level wool-processing activities were established under the control of the SMC. Chapter 9 details the many potential problems these local processing activities face. The move by the county SMCs into wool processing has altered their incentives and especially their response to more open markets. Specifically, as a manufacturer rather than just a handling agent, county SMCs now have a strong incentive to provide their plants with guaranteed raw-wool supplies at low prices. County governments, either through their direct investment in these wool-processing activities or indirectly through the potential these activities have to generate fiscal revenue, also have an interest in guaranteed low-cost supplies to these plants.

The close links which exist between the county SMC and the county government often mean it is difficult to separate their respective interests. At the local level, therefore, the SMCs are likely to receive preferential treatment and encouragement from the authorities irrespective of what national, provincial or even prefectural policies might say about the existence of free markets for products such as wool.

7.2.3 Prefectural SMCs

In many cases, prefectural SMCs are not involved directly in the physical handling of wool but rather they are concerned with the organisation of its distribution. Thus the prefectural SMC informs or sanctions the Animal By-Products Company at the county level to send wool direct to a particular processor. That is, the prefectural SMC has traditionally operated more as an administrative arm of the government than as part of the marketing network of the SMC. This has been explicitly recognised in some cases where in the past the prefectural SMC has received funding not from the regulated SMC marketing margin but directly from the prefectural-level administration.

The extent of the administration task facing a prefectural SMC in a wool-producing area is illustrated by the case of Chifeng City Prefecture SMC in IMAR which has 203 basic SMCs operating at the township level, which in turn operate 756 branch shops. A county SMC operates in each of the 12 counties in the prefecture and there are a total of 62 specialist companies at the county level. Each of the 12 county SMCs has an animal by-products company. The total value of turnover for the prefectural SMC, which has four prefectural-level specialist companies associated with it, was ¥1.3 billion in 1988.

Many of the wool textile mills, especially those up-country mills in the pastoral region, operate at the prefectural level and so have close contacts with the prefectural-level administrations and SMCs. Consequently, much of the wool in the pastoral prefectures was traditionally reserved for the prefectural-level mills. For instance, in the days when wool was part of the planned economy, about 90% of the wool purchased by the Chifeng City SMC was sold to the six prefectural-level wool processors. Under the regulated distribution system in place at the time, the prefectural government procurement quota was 8,120 tonne in 1989 compared with the prefectural production of 10,500 tonne. The development of county scouring facilities and the opening of the market have placed pressure on this traditional distribution pattern as outlined in Chapter 9. Instead of acting as a handling agent sending raw wool direct to the prefectural SMC, many of the county SMCs now retain and scour their own wool and sell it direct to the prefectural mills or to mills elsewhere in China.

Other up-country mills face similar difficulties in retaining their traditional sources of raw wool. Before 1992, the Cabucaer County quota was around 700 tonne of greasy wool. However, there have been no quotas since that year. In Cabucaer County, all the wool purchased by the county SMC has traditionally been sold to the prefectural-level mills in the prefectural capital, Yining. The total amount of greasy wool purchased each year by these mills is around 8,000 tonne, while Yili Prefecture produces about 9,000 tonne of raw wool per annum. The Yili prefectural mills have a captive market, as transport to other mills in XUAR is difficult and the mills in neighbouring Kazakhstan cannot pay in cash. Nevertheless, the mills may find it increasingly difficult to obtain sufficient raw materials if farmers in Cabucaer County and the other counties in Yili Prefecture continue to switch out of fine wool sheep and into meat sheep in response to the growing market and good prices being offered for mutton for export to Kazakhstan.

If the SMC organisation is to be restructured following the emergence of more open wool markets in China, then the units under greatest threat of being made redundant are the prefectural-level SMCs. Many areas currently operate without a prefectural-level SMC, and in those areas where one does operate, many mills and county-level SMCs would prefer to negotiate directly with each other, thus bypassing the prefectural SMC. Mills regard the marketing margin paid to prefectural SMCs (irrespective of whether they contribute marketing services or not) to be an unnecessary addition to the price of their raw-wool input. Other service roles currently performed by the prefectural SMC such as the training of wool-grading technicians could be provided by provincial- or county-level units or by other independent agencies. Consequently, if prefectural-level SMCs are to survive, they will need to improve their marketing services or find particular niche roles, such as being a favourably-located transshipment point with special knowledge about the wool market and good contacts with major mills.

7.2.4 Provincial SMCs

The provincial-level SMCs are even less involved with the handling or processing of wool than the prefectural SMCs. Much of their work is in the policy area, either coordinating the implementation and monitoring of particular wool-marketing regulations or interacting with the provincial governments, many of which are in autonomous regions and so enjoy wide discretionary powers. One important role in the 1989 to 1992 period was the administration of wool stocks. This and other administrative functions of the provincial SMCs are discussed further in Section 7.3.

At the provincial level, the SMC-controlled animal by-product and native goods companies have always had links with separate animal by-product companies within MOFERT. These MOFERT companies, which are actively engaged in foreign trade involving animal by-products, exist in XUAR, IMAR, Gansu, Hubei, Ningxia and several other provinces. Apart from their external trade activities, they differ from the similarly-named companies under the SMC in that they have no subsidiaries at the county or grass-roots level. The MOFERT animal by-product companies have been involved particularly with the purchase and sale of carpet wool and carpet-wool products. They can also purchase wool from their SMC namesakes by direct negotiation. Although there was never any compulsion for SMC companies to sell to the foreign trade companies, under the traditional unified purchase and distribution system a part of the wool purchase quota was allocated for these purposes.

The Tea and Animal By-Products Company in XUAR is under the provincial SMC. This company has about 1,000 staff at the provincial level. There are branches of the company in each of 15 prefectures and prefectural-level cities and in 83 counties. The company operates one wool-processing factory in Urumqi as well as a hide-processing factory and an artificial-leather-making factory.

The Wool and Cashmere Factory is another kind of commercial enterprise under the provincial SMC in XUAR. The factory buys up to 9,500 tonne of raw wool and cashmere each year, consisting of around 6,000 tonne of fine wool, 1,000 tonne of improved wool, 1,500 tonne of local wool and a variable amount of cashmere, depending on the market conditions. Usually most of this wool is sold to textile mills in other provinces, either as raw or scoured wool or after it has been made into tops. The Wool and Cashmere Factory can scour about 5,000 tonne of greasy wool per year, and it has the capacity to process about 700 tonne of clean wool annually through to the top stage.

In Gansu Province, the SMC network comprises 14 prefectural-level units, 86 county cooperatives and 961 cooperatives at the grass-roots level. The whole cooperative network consists of over 15,000 business units, and around 85% of all rural households have shares in the grass-roots cooperatives. The provincial SMC itself operates 37 companies and units. Of these, 27 are business companies and the remaining 10 are institutions of various kinds, including an SMC school, two SMC cadre colleges, a science and technology research institute, an economic research unit, a computer centre, a newspaper agency, a financial services institution and a supply and marketing economics association. The SMCs in Gansu purchase animal products worth over ¥100 million from farmers each year. Wool is the major product handled by the SMC, however, with wool purchases accounting for 70 to 80% of the total amount paid to farmers.

In IMAR, cashmere and wool represent 80% of the business of the Animal By-Products Company operated by the provincial-level SMC. Overall, there are more than 2,000 grass-roots SMC buying stations for cashmere and wool in IMAR.

All provincial SMCs pay an annual membership fee to the All China (National) Union of SMCs. These funds are used to assist SMCs in poor provinces. For instance, the Gansu provincial SMC pays an annual membership fee of ¥200,000 but, as Gansu is a poor province, the provincial SMC receives back almost ¥700,000.

7.3 Administrative and Non-Marketing Role of SMCs

The SMCs display a close link with the various levels of government and have often been used as the institutional vehicle for achieving particular government objectives. The political rather than commercial-marketing dimension of SMC activities was lucidly illustrated in the late 1980s when the SMC began to accumulate large stocks of wool as a result of the policy decision by provincial governments to maintain relatively high State prices for the lower-quality wool produced primarily by individual herders. Nationally, the SMC stockpile of greasy wool rose sharply from 68,000 tonne at the end of 1988 to 120,000 tonne a year later and 143,000 tonne at the end of 1990 (Lin, 1993). Stocks of raw wool held by the SMC peaked in the IMAR in 1990 at around 68,000 tonne, fell in 1991 to 30,000 tonne and rose again in 1992 to 40,000 tonne. At various times during the 1989 to 1992 period, wool-stock levels in the IMAR were more than 20,000 tonne in excess of "normal" levels. Most of the wool in the stockpile was fine wool Grade II and improved fine wool Grade I. The rapid build-up of stocks in 1989 was a function of the collapse of the wool market in 1989 following the buoyant "wool wars" period, the regulated marketing channels and the inability or unwillingness of provincial governments to lower the high purchase prices for raw wool, especially the prices for some of the lower-quality types for which there was little demand.

The situation in XUAR, the other major fine wool-growing province, between 1989 and 1992 was similar to that in IMAR. For example, the SMC in XUAR held about 10,000 tonne of greasy wool in stock at the end of 1991. This wool was valued at over ¥100 million.

The costs associated with these wool stockpiles caused large operating losses for the SMC and put great pressure on the organisation. Furthermore, at least part of these stocks was purchased well before the 1989 slump during the height of the "wool wars" at very high prices. The State was forced to subsidise the holding of wool stocks by the SMC between 1989 and 1992. The old Ministry of Commerce provided ¥500 million for the stocks; the funds were provided as an interest-free loan administered by the prefectural SMC and with no fixed repayment period. The township SMCs were also provided with special loans through the Agricultural Bank of China so that they could keep purchasing wool from the farmers in 1989, 1990 and 1991 despite having large amounts of capital tied up in stocks.

In future, with a free market, the SMCs should be able to avoid an excessive build-up of stocks of any particular grade of wool by allowing the market to determine the price for various grades. However, as discussed in Chapter 5, the prices that emerge from the free market may be unacceptable to the government from an income distribution perspective in the longer term as it is the poorer households that tend to produce the lower-quality wools. It was primarily these lower grades of wool which made up the SMC stockpile in the 1989 to 1992 slump. Should provincial governments decide to support wool prices or to moderate the grade-price differentials generated in an open market, they will need to intervene through purchasing and the most likely vehicle for this will be the SMCs. The possibility that the SMCs will be required to play a future price-support role could seriously jeopardise the chances that the SMCs will become more or less independent publicly-owned business enterprises (corporations) or revert to farmer-owned cooperatives.

7.4 Breaking the SMC Monopoly

As the data in Fig. 7.1 would suggest, the SMC began to experience some competition in relation to its wool-purchasing business in the early 1980s. By 1984, it had lost almost 10% of the total clip to other marketing channels of which the two most important were direct purchases from State farms by mills and the wool-buying activities of other government units. These transactions outside the official SMC channel were presumably sanctioned on the basis that they involved over-quota wool.

The wool-marketing reforms of 1985 enabled these non-SMC buyers, together with many other new entrants, to cut the SMC share of the wool trade to just over 60% in that year. The SMCs fought back, pushing their market share back up to almost 80% in 1988 and 1989. They achieved the improved result in 1988 by out-competing other buyers on a price basis, a strategy which left the SMCs holding large stockpiles of grossly over-priced wool when the market collapsed in 1989. In 1989 the high market share was achieved largely because the SMCs became the "buyer of last resort". In 1992 and 1993, once the wool market was re-opened the SMCs saw their market share fall rapidly.

This section examines the development of alternative marketing channels and discusses the responses of the SMCs to the breaking of their monopoly.

7.4.1 State Farms and Direct Sales

State farms are an important part of the wool production sector in China. Since the economic reforms of the late 1970s, State farms have often been treated differently to individual households with respect to wool marketing. Indeed, various aspects of the wool-marketing reforms have been "tested" on the State farms. State farms are often better suited than individual herders or grass-roots SMCs to test the marketing reforms as they typically have more homogeneous lots, higher-quality wool and more skilled wool classers. In particular, these characteristics lend themselves to direct sales from the State farms to the mills, thereby avoiding the traditional marketing channels which were costly and inefficient. Apart from reducing the marketing margin, direct selling has a number of other advantages as noted by Lin (1993). These include improving the awareness of the producers of the requirements of the processors and facilitating the channelling of technical advice and possibly investment capital from the mills to the producers.

State farms in IMAR (both ordinary State farms which produce wool and the special State sheep stud farms) have been permitted to sell their wool directly to processors since the 1990 wool-selling season, when all wool production quotas for the State farms at the provincial level were cancelled. In this case, the mills must negotiate their purchases through the provincial-level stud-farm company under the

IMAR Animal Husbandry Bureau rather than through the SMC network. The State farms in the IMAR are said to supply the best wool produced in the province and the mills pay a premium of up to 10% to obtain this wool. Most of the wool purchased from the State farms is said to be fine wool Grades I and II. About 10% of the raw-wool requirements of the mills in Huhehot can be purchased from the State farms, with the remaining 90% coming from the prefectural SMCs. Most of the wool purchased from the SMCs is fine wool Grade II or improved fine wool Grade I.

Large national- or provincial-level State sheep-stud farms such as Huang Cheng in Gansu and Gongnaisi in XUAR, have had the right to sell their wool directly to mills since the mid-1980s. But, in general, lower-level State farms in most provinces had to market their wool through the SMC until the market was declared open in 1992. Often it was the local-level government and not the provincial government which enforced this regulation. For example, while the Gansu provincial government declared a free market for wool in Gansu from 1985, the Sunan County government in Gansu required the State farms in Sunan (except for the provincial-level Huang Cheng stud farm) to sell their wool to the county SMC and that this wool be processed through the county-government-owned scouring plant.

The Aohan stud farm is a prefectural-level State farm in the IMAR, and as such, it is permitted to sell directly to processors. Currently the entire wool clip, something in the order of 80 tonne greasy per annum, is sold directly to processors. With the recent opening of the county wool-scouring plant in Aohan, it will be interesting to see whether the county government will try to coerce the Aohan State Farm into supplying its wool to the county scours.

State farms are of particular importance in XUAR which is one of the two major fine wool-producing provinces in China. The total production in XUAR is about 48,000 tonne per year. About 6,000 tonne of this is grown on State farms under the control of the AHB. The Production and Construction Corps (PCC) State farms produce a further 9,000 tonne per year. Thus around 15,000 tonne out of the total 48,000-tonne XUAR clip is grown on State farms. State farms generally produce better wool and a higher proportion of finer wool than private households. Hence, a high proportion (perhaps over half) of the fine wool produced in XUAR is grown on State farms.

Wool sold by State farms directly to the mills has usually been sold according to the old National Wool Grading Standard on a clean-wool basis. In recent years, about one dozen of the larger, more modern State farms in XUAR have sold their wool direct to mills on a clean-scoured basis according to industrial grades. As mentioned in Section 6.5, the FIB does the testing of this wool and the price paid depends on the clean yield.

While the amount of direct sales varies from province to province, one of the major constraints to further development of this marketing channel has been the inability of mills to provide sufficient buyers. To fill this gap, another trading company known as the Animal Husbandry Industrial and Commercial Company (AHICC), under the control of the Ministry of Agriculture at the national level, has emerged as a go-between for the mills and the State farms. It operates in direct competition with the SMC in the provision of wool-marketing services. The AHICC exists at the province level in all the major animal production areas such as XUAR, IMAR, Gansu, Heilongjiang, Jilin, Liaoning, Shaanxi, Sichuan and Shanxi. Subsidiary companies of the AHICC also exist in some prefectures, and even at the county level in some areas.

The XUAR AHICC, for example, was established in 1980, but did not really commence commercial activities until 1984. Initially, two subsidiary companies were formed: the XUAR AHI Trading Company which has responsibilities for buying and selling animal products (and some agricultural products), and the XUAR AHI Breeding and Livestock Company, which trades in livestock. Further subsidiaries have since been added. In 1986, a subsidiary was established to operate the Tian Shan Hotel in Urumqi, and the Animal Products Trade Centre was set up to provide wholesaling services and to operate a retail department store. In 1988, an Animal Medicine and Machinery Company made up the fifth subsidiary of the AHICC. Additionally, in 1983, the XUAR AHICC entered a joint venture with the national AHICC and the Urumqi Agricultural Committee (an organisation under the Urumqi City government) to form the Hua Xin AHICC, which now operates egg and chicken production enterprises. The XUAR AHICC has a 25% share of this joint venture. The AHICC only operates in two prefectures in XUAR, namely: Altay and Yili. In the absence of the AHICC, the types of services it provides are met by the State farms, the SMC, the Food Company, etc. The XUAR AHICC has purchased raw wool since commencing business in 1984. It has not bought wool from townships or individual households as it has insufficient staff to service this activity. Instead, it buys wool from State farms, and generally only from the larger of the farms under the control of the XUAR AHB. In 1990, for example, the XUAR AHICC purchased around 4,500 tonne or about 75% of the wool produced by the State farms in the XUAR AHB network. It purchases very little wool from PCC State farms.

The State farms often find it convenient to deal through the AHICC rather than selling direct to mills because the Company is in a position of having good information on prices being offered by mills, and of being able to supply mills with large amounts of good wool. As a result, State farms can receive a better net price than if they tried to do their own marketing. The Company has good relations with a wide range of mills. The Company claims that it would be difficult for State farms to make these contacts directly and economically. The AHICC not only has a competitive advantage over the SMC in its institutional affinity with AHB State farms and with the Ministry of Agriculture but it also operates with smaller margins. Whereas the typical SMC margin is in the order of 17%, the corresponding margin for the AHICC is about 15%. However, of this, only 5% is required to cover the cost of handling, with the remaining 10% being a trading profit, of which 6% is eventually returned to the State farms from which wool has been bought. Returning some of the profits to the State farms after the Company actually sells the wool implies a forced loan by the State farms to the AHICC, and in the free market situation where selling prices are not fixed and where sales can be delayed, it amounts to a sharing of the marketing risk.

The AHICC buys on a clean-wool basis and most sales are also on this basis. The Company accepts the grading done on the State farms because the classers are AHB-trained. Testing for clean yield is done by the FIB at county or prefectural level. The Company does no testing of its own, nor does it attempt any regrading, nor repacking from the 70–100kg packs used on the State farms. The State farm bears responsibility for transportation to Urumqi and for the packing and testing fees. The latter were set at ¥0.12 per kg and ¥0.06 per kg respectively in 1992. Transport fees vary with mode of transport and with distance.

Another key institution involved in wool production in the XUAR is the Xinjiang PCC. As mentioned earlier, the PCC State farms account for one-fifth of total wool production in XUAR, and their wool is generally finer and of better quality than the average clip. Unlike the State farms which operate under the AHB of the Xinjiang Agricultural Commission (Ministry of Agriculture), the PCC activities are under the control of the Xinjiang General Bureau of State Farms and Land Reclamation. Apart from the State farms, the Xinjiang PCC also operates wool-processing facilities such as the large fully-integrated mill in Shihezi. PCC systems operate in Yunnan, Hainan and Heilongjiang Provinces as well as XUAR. For a detailed discussion of the PCC, see Longworth and Williamson (1993, pp.148–153).

The PCC State farms traditionally operated under different conditions to those applying to other kinds of State farm or to individual herders outside the State farm system. Before 1986, the Tea and Animal By-Products Company of the SMC purchased all PCC State farm wool, but at the same time, beginning in 1966, PCC farms were allowed to deal directly with mills (principally PCC mills). In the past, the PCC mills were required to accept all wool offered by the PCC farms. This usually represented around 2,000 tonne of greasy wool per annum. With the opening of the wool market in 1992, the PCC mills are no longer under any obligation to purchase PCC farm wool, nor are the PCC farms obliged to supply the PCC mills.

Apart from deals involving the State farms, there have been other experiments with direct sales in recent years. Since 1989, there has been a provincial policy in XUAR allowing textile mills to purchase coloured and native wool direct from herders. However, county SMC officials (in Hebukesaier County for example) claimed that mills took the (illegal) opportunity to purchase good wool from herders at the same time. Apart from avoiding the SMC margin, the mills also avoided paying the wool product tax, according to the SMC officials. The loss of business to direct sales was of major concern to the SMCs in Hebukesaier County (see Section 7.4.3).

Another modest reform involved eliminating the role of the prefectural-level SMC in wool marketing by allowing the textile mills to buy directly from county-level SMCs. Indeed, in some areas there may be considerable advantages in having the mills buy directly from the grass-roots or township-level SMCs, thus eliminating both the prefectural- and the county-level SMCs from the marketing chain.

7.4.2 Other New Marketing Channels

Direct deals between mills and State farms, county SMCs, township SMCs or even individual herders are but one change to the regulated SMC marketing channels encouraged by the wool-marketing reforms since the mid-1980s. Another key change has been the introduction and development of wool auctions. To date, only rather rudimentary auctions accounting for a minor share of the total wool marketed each year have emerged. In the longer term, and with further development, however, they offer considerable scope for improvement in wool marketing, and their potential importance warrants separate discussion in the following chapter.

Apart from auctions, other forms of centralised selling are emerging. For example, a Wool Trade Fair was held in Huhehot in 1991. The fair was organised by the State Production Commission under the State Council, along with the AHB, the Textile

Industry Corporation, the State Price Bureau, and the Fibre Inspection Bureau. The sellers were SMCs (more than 200) from various provinces, while the 300 buyers came from 32 provinces and cities. The fair sold 14,000 tonne of wool, of which 60% was fine, 35% was improved and 5% was other wool. Unlike the auction system, no stringent conditions were imposed on quality. The Wool Trade Fair in Huhehot was the first such specialised wool trade fair ever held in China. Other kinds of trade fairs are fairly common. For example, each year there is a mixed-produce trade fair in Hunan Province and a much smaller trade fair run by the SMC in Zhengzhou, the capital of Henan Province. The Planning Committee authorised the specialised fair in Huhehot, and its main purpose was to reduce the SMC wool stockpile. While 14,000 tonne was sold, almost 70,000 tonne was offered for sale. One of the reasons for the poor clearance was that the fair was held in May and most purchasers were prepared to wait for new-season wool which would become available from early June. The sale was by sample, with inspection certificates issued by the FIB.

Watson (1988) outlines the growth in wholesale markets in agricultural products which occurred in China throughout the 1980s. He argues that the rapid and widespread emergence of wholesale markets proceeded according to its own economic logic and was the inevitable result of the introduction of the household production responsibility system and price reforms. Wholesale markets for wool, however, have not been as quick to emerge. The AHB in XUAR had plans to establish a wool wholesale market in Urumqi in 1993. Private wool traders were to be encouraged to sell in the planned wool wholesale market. Traders will be required to pay a 1% commission on their total sales value to the AHB organisation controlling the wholesale market. Warehouses in Huhehot and Nanjing were also gearing up for "sale by sample" experiments in the early 1990s.

7.4.3 SMC Responses to the Opening of the Wool Market

By 1992, the official move towards re-opening the wool markets was considered an inevitable trend that even the SMCs could not stop. SMC (Animal By-Product Company) officials in IMAR considered there would be a negative impact on the SMCs initially but not in the longer term. The SMCs had a number of important advantages over other potential wool-buying agencies. In particular, the vast network of SMC purchase stations backed by well-established handling and storage facilities could not be easily matched. In addition, the SMCs had much better access to finance and had a big pool of people with wool expertise. In summary, the SMCs were well placed to remain the dominant force in the domestic wool trade after they adjusted to the realities of competition in a free market. Overall, the SMCs throughout the pastoral region seemed much more ready to be pro-active in 1992 than in 1986 at the start of the "wool wars".

County-SMC managers in IMAR met in March 1992 to discuss the developments. Among other strategies, they decided to adhere to the principle of quality first. That is, they agreed to abandon welfare considerations and accept commercial realities in relation to wool. Putting a strictly commercial buying policy in place will entail implementing a number of important new approaches. First, responsibility for wool quality lies directly with the individual buyers operating at the purchasing-station level. If their grading is incorrect, it was decided that they would be fined. The fine would be determined in consultation with the FIB. In addition, the issuing of certificates to people who had been trained as wool technicians would be tightened by the relevant ministries and the FIB. Second, SMC buyers would be allowed to refuse to purchase wool of low quality. If the wool was too dirty, herders would first have to clean it. Third, in future there would be no State or fixed prices. Each SMC unit would be free to make its own commercial decision about how much to pay for wool.

County and township SMCs also intended to maintain business by offering better services to herders: by providing better-quality shears and other tools and by stocking veterinary medicines; by providing more convenient ways of collecting the wool from individual herders (perhaps the greatest incentive for herders to deal with the SMC); by providing accommodation and other conveniences for herders coming to their offices; and by engaging in some barter trade such as the provision of fertiliser in exchange for wool. The SMC managers also felt they could depend upon having significant financial backing through the Agricultural Bank of China and their own accumulated funds. They were also of the view that the county governments would support the SMCs both by providing access to loans and by new wool product tax arrangements which shifted the responsibility for paying the tax back on to the herder (or State farm) growing the wool.

The SMC management in Wushen County, for example, intended to compete with "private buyers" by reducing their traditional 17% marketing fee and paying the farmers higher prices. Furthermore, Wushen is in the Yikezhao Prefecture of IMAR and the Yikezhao prefectural administration has introduced a special tax of 9% to be paid by all private (non-SMC) buyers of wool. The 9% prefectural levy is made up of: 3% pasture construction fee (paid into the Pasture Construction Fund); 2% market management fee; and 4% fee for rental of space, use of public facilities, etc. In this case, it would appear that the prefectural government has deliberately introduced a new tax designed to protect the SMCs. The Wushen SMC had already signed contracts in early June 1992 (before the start of the main shearing and purchasing season) for the sale of Wushen wool to Nanjing and Lanzhou topmaking mills, and to the Animal By-Products Company in Huhehot (provincial level), the Animal By-Products Company in Yikezhao Prefecture (prefectural level), and the Animal By-Products Company in Wuhai City (county/city level). These contracts covered the total expected wool production from the county in 1992. Clearly, the Wushen SMC was not expecting private buyers to be able to compete for Wushen wool in 1992. Furthermore, the Wushen County SMC was also planning to take advantage of the open market to buy selected Erdos fine wool from the other three pastoral counties in Yikezhao Prefecture and market it as if it came from Wushen County. This is because Wushen County has developed a reputation for high-quality wool. That is, the Wushen SMC were planning to expand their commercial operations and sell much more wool than was produced in Wushen County. From their viewpoint, the open market provided new opportunities. They emphasised that in future they must "police" the buying from farmers to make sure they satisfied their contractual obligations as regards quality. They intended to buy only the best-quality wool from the neighbouring counties so that they could maintain the Wushen County "image" for quality.

Hebukesaier County SMC in XUAR faced an entirely different situation to Wushen County when the market re-opened in 1992. Nevertheless, it also had the same desire to maintain or increase its business. As indicated earlier, the Hebukesaier SMC had progressively lost business in the lead up to the opening of the wool market in 1992. In 1989, it purchased 360 tonne or almost two-thirds of the wool produced in the county. However, its purchases fell to 197 tonne in 1990 and 180 tonne in 1991. The lost business was not only the result of direct sales to mills by the four State farms in the county (which produced a total of 250 tonne of raw wool per year) but also direct sales to mills by herders. The immediate problem facing the Hebukesaier County SMC was that it had a staff of 120 and it was finding it difficult to survive on the margins from the reduced quantities of wool purchased. With the opening of the market in 1992, the SMC expected its purchases to drop even further unless something was done.

Apparently, discussions were held between the SMC and the four State farms about purchasing their wool. All four farms agreed to sell their wool to the SMC in 1992 based on 1991 official State prices plus a premium of 21% (which was the premium paid by textile mills in 1991 for State farm wool). However, given the market prices for wool in 1992, the SMC would have had extreme difficulty in selling the State farm wool to mills at a profit. The Hebukesaier County SMC claimed in 1992 that the provincial-level SMC was underwriting its purchases from the State farms as well as the purchases of herders' wool for which the county SMC was offering the same price as in 1991. In an effort to maintain purchases from herders, Hebukesaier SMC introduced "roving" purchase stations in the 1992 season. The contract with the State farms and the more convenient "roving" purchase stations enabled the Hebukesaier SMC to purchase around 450 tonne of wool in 1992, which was well over double the amount purchased in 1991. However, it is doubtful that it could have achieved this increased market share without making a large trading loss. Even if the provincial SMC was prepared to subsidise the county operation in 1992, such an arrangement is unlikely to continue in the longer term.

With the opening of the wool market, everyone wishing to purchase wool must obtain a licence to do so from the Industrial and Commercial Administration Bureau, an administrative unit directly under the State Council. It is understood that these licences attract a substantial fee and will only be issued to individuals who can demonstrate that they have been properly trained in the new National Wool Grading Standard. Furthermore, it seems a separate licence is needed for each county. Clearly, the administration of these licences at the county level provides another means by which local authorities can "bend the rules" in favour of the SMCs.

The two examples just discussed, Wushen and Hebukesaier, demonstrate the different ways the local SMCs have responded to the opening of the wool market and the extent to which the authorities are prepared to go when necessary to assist the SMCs. Despite a more pro-active stance, in general the SMC wool business has been markedly affected by the opening of the wool market. In the IMAR, total SMC purchases in 1992 were down by about one-quarter relative to 1991 and accounted for only 54% of wool produced in the autonomous region in 1992. However, the level of SMC purchases varied from county to county. In Wushen, the SMC purchased almost all the wool. In Alukeerqin County, the SMC purchased about two-thirds of the 1,500 tonne of wool produced in the county in 1992. The remaining one-third was mainly purchased by other State-owned companies operating in the county and some township enterprises. In addition, buyers came from outside the county, including buyers from Shandong and Hebei Provinces. In neighbouring Balinyou County, the opening of the market had a much greater effect on SMC purchases. Typically the SMC purchased around 900 to 1,000 tonne each year, but in 1992 it purchased only

300 tonne, representing only one-quarter of the wool produced in the county. Other buyers included the food company, the foreign trade company, and some other government units as well as outside buyers.

For China as a whole, Fig. 7.1 demonstrates the sharp decline in the proportion of the national clip purchased by the SMCs in 1992 and 1993.

7.5 Concluding Remarks

The SMCs are at a watershed in their 40-year history. Can they adapt and build upon their enormous distribution network and market infrastructure to remain a major force in the more open Chinese wool markets or will the legacy of past cosseting see the demise of this agribusiness empire as a result of the new competitive environment? Local governments are likely to continue to provide some support for the SMCs, because at the local level they are seen as a convenient medium for the generation of fiscal revenue and employment and for pursuing certain social policy objectives. However, the local support is not unlimited, nor is it likely to be sustained at high levels for extended periods of time. SMCs will have to adjust to the new commercial realities if they are to remain a major force in the Chinese wool market.

The message from this chapter is that to survive and develop, SMCs will have to grasp the opportunities being created by the new agribusiness reforms. One such opportunity was identified in Section 6.3 where it was pointed out that township or county SMCs could become "wool merchants" by interlotting and regrading wool. The enormous network of potential service points and their influential contacts with local authorities are the two main strengths of the SMCs. These advantages should make the SMCs most attractive joint-venture partners for foreign agribusiness entrepreneurs wishing to participate in the modernisation of the Chinese rural sector to mutual advantage.

Chapter 8

Auctioning the Lot

Among the more innovative elements of Chinese wool-marketing reform in recent years has been the introduction of wool auctions. Auctions have the potential to improve price determination, bring buyers and sellers closer together and focus attention on issues such as grading, objective measurement and quality improvement.

Most of the raw wool in the world passes through auctions and most of the major wool-producing countries have well-developed auction systems. However, despite an apparently successful introduction in the late 1980s, wool auctions in China subsequently foundered in the 1990s. An examination of the development of the auction system since the late 1980s provides some fascinating insights into the way Chinese decision-makers have approached this particular agribusiness reform.

The evolution of wool auctions is briefly outlined and placed in the context of overall wool-marketing reform in China in the first section of this chapter. The special rules governing the auctions including the standards applying to wool sold at auction are then described. The body of the chapter is focused on two key auctions held in 1991. These two 1991 auctions were critical in that they immediately followed the introduction of new regulations governing the operation of the auctions and they were two of the last three auctions held before the deregulation of the wool markets in 1992. Information appearing in the catalogues for these sales is presented and an analysis is made of the prices paid at these auctions.

Chinese wool auctions have virtually been suspended since 1991 despite an improvement in wool markets and the more liberal marketing environment. Some underlying reasons for the demise of the auctions and the steps that need to be taken if auctions are to realise their full potential as an agribusiness reform are discussed in the last section.

8.1 Evolution of Wool Auctions

Auctions arose during the height of the "wool wars" in the second half of the 1980s (Zhang, 1993). At this time, discerning buyers were seeking alternative purchase channels and improved quality, conditions just right for the development of auctions. Both the MOA and MOTI were keen to push for the development of auctions as an alternative marketing outlet to the State-controlled monopoly channels through the SMCs and the Ministry of Commerce.

The decentralisation of wool-marketing policy in the mid-1980s and the subsequent "wool wars" had encouraged alternative buyers but the resultant chaotic markets brought about a lowering of wool quality. The Ministry of Agriculture saw auctions as one means of raising the profile of wool and wool marketing in China and as a means of preventing the "wool wars" from derailing its long-term strategies for improving the quality of domestic wool. For MOTI, the auctions provided a means of identifying and obtaining the best-quality domestic wool from the remote producing regions for use in its large, export-oriented, east coast mills. The chaos of the "wool wars" encouraged the Ministry of Agriculture and MOTI to join forces.

In 1986, the Ministry of Agriculture and MOTI jointly established a Leading Group at the Central government level to develop the "wool production base" of China. Subsequently, each interested province set up its own "leading group" and developed its own "wool base" concept. Beginning in 1987, the Leading Group in the XUAR invited 15 State farms (including PCC farms) well known for the quality of their wool to participate in a wool auction in Urumqi (the capital of XUAR). This first experimental auction involved only 78 tonne of greasy wool (Table 8.1). Although of small volume, the initial auction was well received and was followed by a series of larger auctions in 1988 in Urumqi, Huhehot, Jilin and Nanjing. Combined offerings amounted to almost 1,500 tonne of raw wool, all of which was sold in the buoyant wool market of 1988.

Following the success of the 1988 auctions, it was decided to hold a large national auction in Beijing in August 1989. The 3,100 tonne of greasy wool offered at the auction in Beijing was more than double the total sales of the previous year. However, as discussed in Section 3.4, by August 1989 the wool boom was over. The depressed market conditions saw less than half of the offerings sold, and at considerably reduced prices compared with those received in 1988. All the wool on offer, nevertheless, was sold either at the auction or in post-auction negotiations.

In 1990, the auctions returned to the main wool-growing regions. Volumes offered in 1990 at the auctions held at Huhehot (782 tonne) and Urumqi (917 tonne) were much higher than the quantities offered at these centres in 1988. But demand was weak and the prices paid for the wool sold under the hammer were below those achieved at the 1989 Beijing sale and almost half the wool had to be sold following the auction. While the total volume of wool offered at the three auctions conducted in Huhehot, Xi'an and Qiqihaer in 1991 increased, it did not reach the level achieved at the single Beijing sale of 1989. The proportion of the wool on offer which was sold at auction (the "clearance rate") was a little higher in 1991 than in 1990 but prices were much lower (Table 8.1).

In August 1992, wool auctions were held in Huhehot (390 tonne clean wool offered for sale) and in Nanjing (400 tonne clean wool). The combined offering, which was equivalent to around 2,000 tonne of greasy wool, was well down on the offerings in the previous year. Remarkably, however, none of the wool offered at these auctions was sold. The situation deteriorated in 1993. Despite plans to hold one or two wool auctions, the intended sales had to be cancelled owing to a lack of sellers prepared to offer wool for inclusion in the auctions (Zhang and Niu, 1994).

The auctions usually have been held in the main wool-producing regions, although some auctions have been conducted in major processing centres (Nanjing and Beijing). In general, the auctions have been organised by the Textile Industry Corporations (TICs) in the particular province in which the auction is held. The fledgling auction marketing channel, when it operated, has been restricted to handling mainly fine and improved fine wool. The peak offering of 3,100 tonne (greasy wool) in 1989 represented about 3% of this type of wool grown in China in that year. However, the wool marketed through the auction channel would have represented a much more significant share of the best wool produced in China.

Table 8.1 Details of All Wool Auctions Held in China Prior to 1994

Year	Location of auction	Amount of wool involved				Average price paid at auction
		Greasy-wool basis		Clean-wool basis		
		Offered	Sold	Offered	Sold	
		(t)	(t)	(t)	(t)	(¥/kg clean wool)
1987	Urumqi	78.2	78.2	46.3	46.3	34.45
1988	Urumqi	677	677			na
	Huhehot	326.5	326.5			63.12*
	Jilin	104.6	104.6			na
	Nanjing	385.7	385.7			na
	Sub-total	1,493.8	1,493.8			> 60
1989	Beijing	3,124.6	1,516.0	1,927.8	938.6	39.37 (36.43–43.9)
1990	Urumqi	916.5	511.3	487.2	270.6	31.11
	Huhehot	781.6	437.6	337.9	189.2	31.29
	Sub-total	1,698.1	948.9	825.1	459.8	31.15
1991	Qiqihaer	990	940	416	399	(22.1–32.0)
	Xi'an	993	597	491	295	(21.3–32.8)
	Huhehot	773	591	332	252	(22.9–33.5)
	Sub-total	2,756	2,128	1,239	946	around 30
1992	Huhehot			390	Nil	
	Nanjing			400	Nil	
	Sub-total			790	Nil	
1993	No auctions held					

na, Not available.
*This price was paid for 264.1 tonne greasy wool (equivalent to 146.3 tonne clean wool) which came from Xinjiang. The overall average prices received at the 1988 sales are not available.
Source: Zhang and Niu (1994).

8.2 Wool Auction Standard and Rules

To understand why Chinese wool auctions have evolved in the manner outlined above requires a closer examination of the way they have been conducted and the rules under which they have operated. A draft "Work Regulations for Wool Auction", designed to solicit industry opinions, was drawn up by MOTI, the Ministry of Agriculture and other relevant departments in 1989. Subsequently, the Working Group of Wool Auction Reformation Pilots, formally set up in April 1990, redrafted the regulations into a revised edition in May 1990. The regulations were further revised, enlarged and finally renamed "Quality Standard of Auctioned Wool" in March 1991. Since these regulations are not available in English, a translation has been reproduced in Appendix H. The institutions involved in drafting the 1991 regulations represented a broad cross-section of stakeholders in the development of the auction system. They included Central government organisations (Production Commission under the State Council, MOTI, MOA and the National FIB), major provincial research institutes (Shanghai Academy of Textile Sciences and Xinjiang Wool Research Institute), various textile industry corporations in the producing regions (Xinjiang Textile Industry Corporation, IMAR Textile Industry Corporation), production-oriented organisations (IMAR AHB Animal Improvement Station), and major organisations representing the mills (Nanjing Branch of China Textile Goods and Material Company, Department of Raw Materials of the Jiangsu Textile Industry Company and Shanghai Wool Top Company).

The Standard consists primarily of two parts. The first part outlines the quality standards for wool entering the auction, while the second part outlines rules and procedures associated with the auction.

8.2.1 Quality Standards

Since March 1991, wool offered for sale through the auctions must be graded according to the Quality Standard of Auctioned Wool. The quality standard is listed in full in Part A of Appendix H while a summary is provided in Table 8.2. The wool is first divided into homogeneous wool (fine and semi-fine wool) and heterogeneous wool (improved wool). Improved wool is classified into three groups according to the percentage of coarse-cavity wool. The basis for classification of homogeneous wool is

Table 8.2 Summary of "Quality Standard of Auctioned Wool"

	Homogeneous wool		Heterogeneous wool
	Fine wool	Semi-fine wool	Improved wool
Fineness	count (µm)	count (µm)	(% of coarse-cavity wool)
	70 (18.1–20.0)	58 (25.1–27.0)	I (1.0)
	66 (20.1–21.5)	56 (27.1–29.5)	II (3.5–4.0)
	64 (21.6–23.0)	50 (29.6–32.5)	III (6.0–7.0)
	60 (23.1–25.0)	48 (32.6–35.5)	
		46 (35.6–38.5)	
		44 (38.6–42.0)	
Staple length	(cm)	(cm)	
	A (≥ 8.0)	A (≥ 10.0)	
	B (6.0–7.9)	B (7.0–9.9)	
	C (4.0–5.9)	C (4.0–6.9)	
	D (< 4.0)		
Quality grade (according to clip quality and apparent morphology)			
	I		I
	II		II
	III		III

Source: Appendix H: Quality Standard of Auctioned Wool.

fineness, staple length and, in the case of fine wool, quality grade. As can be seen in Table 8.2, fine wool is sorted among four counts or micron ranges while semi-fine wool is sorted among six such categories. The wool is further grouped according to staple length, with fine wool split into four staple-length ranges and semi-fine wool into three. In the case of fine wool Group B (staple length from 6 to 7.9cm) and Group C (4 to 5.9cm), each is further divided into two sub-groups for the purposes of specifying quality differential ratios (QDRs) used in the specification of start prices for the auctions. Finally, fine and improved wool are further classified by quality grade (i.e. according to clip quality and apparent morphology).

The regulations contain a great deal of information about how the wool is to be prepared for sale by auction; about how the wool is to be appraised to determine the grade; and how it is to be objectively measured, etc. For full details see Appendix H.

8.2.2 Auction Rules

Not only does the wool sold at auctions have to be graded according to a specific set of standards but key wool attributes used in the grading (such as micron and staple length) or in information presented in the catalogues have to be objectively measured. Before the 1990 auctions and the draft regulations, testing was carried out prior to auctions if the sellers requested it, or after auctions if there was a dispute over quality. Not all sellers at auctions requested testing, owing to the cost which was set at 0.4% of the value of the wool sold. The seller did not have to pay for the testing until the wool was sold.

Objective measurement of the wool is undertaken by the provincial FIB associated with the supplier in accordance with the quality standards (Part A) and the fibre testing methods (Part D) listed in Appendix H. The objective measurement part of the Standard outlines in detail the sampling methods to be employed, the technical details and specifications of the measurements, the calculations used, and the precision to be reported in the final estimates. The technical acceptance range is ±3% for fineness and ±0.5cm for staple length. Other certification procedures are outlined in Part F(5) of the Standard. The purchasers have 45 days from receipt of the wool to file a claim if they consider the wool does not meet the objective measurements listed in the catalogue, or if other attributes of the wool not objectively measured, such as the proportion of tender wool, are not consistent with the sample provided. Procedures, responsibilities and penalties associated with arbitration of claims are reported in Part F(6) of the Standard.

The auction rules also set out other procedures for the handling of wool coming through the auctions. This includes details regarding shearing and clip preparation (Part B) and rules for packing of the wool (Part E) which specify, among other things, the material from which the wool bales are to be made, and the dimension and net weight of a bale. Each wool bale must be marked with the State farm name, auction number, grade, wool-classer number, gross weight and bale number. The auction number is a six-digit code indicating year, country-region code and lot number. A code letter is added at the end of the number to indicate non-homogeneous wool. Storage and transport rules are also specified in Part E and concentrate mainly on ensuring standard handling procedures across wool batches.

Buyers at the auctions can only be wool-using units (such as mills or their agents), as it is illegal to purchase wool for speculation or intermediate trade. Prior to

1992, buyers needed only to show that they represented an organisation permitted to purchase wool at the auction and that they had some knowledge about wool in order to obtain a permission card to participate at the auctions. The system was tightened from 1992 so that buyers had to register and have a Qualification Certificate of Wool Buyer which demonstrated that they had undertaken a special one-month training course at the Wuxi Training Centre for Modern Management of Wool in Jiangsu Province (Section 6.3.1). Organisations permitted to send qualified buyers to the auctions were large (>10,000 worsted spindles or >5,000 woollen spindles) and medium (>5,000 worsted or >2,000 woollen spindles) textile enterprises and the provincial TICs. Conversely, township enterprises and SMCs were not permitted to send buyers and had to buy wool from the auctions through the TICs.

At the Beijing auction in September 1989, immediate payment in cash was required for all purchases. This proved to be a serious constraint for a number of mills/buyers. Consequently, the 1991 regulations required only that a 30% deposit be paid within 15 days of the auction. Sellers, on receipt of this deposit, then had 45 days to ship the wool. The remaining payment had to be made by the buyer within 10 days of shipment. Any delays in payments or shipments incurred a penalty (set in the regulations at 1.5% interest per month except when alternative arrangements were negotiated).

8.2.3 Price Discovery

Bidding for each lot of wool is on a clean-scoured basis and proceeds from a start price. A start price for the reference wool Grade 64B2II (i.e. 64 count, B staple length sub-group 2, quality Grade II—see Table 8.2) is determined by the local government and price bureau in the province holding the auction, nominally with reference to the wool-marketing situation at the time of the auction. QDRs for other grades of the wool, specified in the regulations, are then used to determine the start prices for other grades of wool given the basic start price for 64B2II wool. If there are no bids at the start price, then the auctioneer negotiates with the supplier for an adjustment to the start price. The institution organising the auction must also set a reserve price for the wool: this was temporarily specified in the 1991 regulations as 90% of the start price.

Thus, two key elements influence price discovery at the auctions, namely the existence of a reserve price and of fixed quality (price) differentials across grades which determine the start prices. Section 8.3 examines this influence by analysing actual auction data for the auctions immediately following the implementation of the 1991 Standard.

The QDRs, setting out the grade-price differentials to be used in the setting of the start prices for each grade, are listed in part F(3) of Appendix H and are reproduced in Table 8.3. As indicated, QDRs are specified for both fine wool and improved wool. For fine wool, QDRs are given for each of the grading criteria, namely fineness, staple length and quality grade, while QDRs for improved wool are specified only according to the quality grades. The QDR for any particular type of wool is obtained by the product of the QDR for each criterion. For example, the QDR for 70B1III wool of 109 is equal to (1.08 [QDR for 70 count wool] × 1.04 [QDR for staple length sub-group B1] × 0.97 [QDR for quality Grade III]) × 100. The QDRs for the 72 different fine wool types (four fineness classes by six staple length classes by three quality grades) are given in the attachment to Appendix H.

The QDRs as reported in Table 8.3 imply that each 1μm increase in wool fineness decreased the QDR by around 2.3%. Assuming an average clean-wool price for the reference 64B2II grade wool of ¥25 per kg clean, this represents a price fall of around ¥0.58 per kg clean for each 1μm increase in mean fibre diameter. Each step up to a higher-quality grade increases the QDR by 3.5% or prices by ¥0.88 per kg clean. The QDRs do not exhibit a simple linear relationship with staple length but instead exhibit an approximate log-linear relationship where the quality differentials increase rapidly with the low staple lengths but taper off at the high staple lengths. For instance, raising the staple length from 4.5cm to 5.5cm increases the QDR by around 15% or by ¥3.75 per kg clean, whereas a rise in staple length from 6.5 to 7.5cm increases the QDR by only 4% or prices by ¥1 per kg clean.

Table 8.3 Quality Differential Ratios (QDRs) for Fine and Improved Fine Wool at Auctions

	Categories	Technical range	QDR
			(% of reference price)
Fine wool			
Fineness		(μm)	
	70	18.1–20	108
	66	20.1–21.5	104
	64	21.6–23	100
	60	23.1–25	96
Staple length		(cm)	
	A	≥ 8	109
	B1	7–7.9	104
	B2	6–6.9	100
	C1	5–5.9	90
	C2	4–4.9	75
	D	< 4	55
Quality grade			
	I		104
	II		100
	III		97
Improved wool			
Quality grade			
	I		86
	II		62
	III		48
	Below III		33

Source: Appendix H: Quality Standard of Auctioned Wool.

The QDRs in the Auction Standard refer to much more clearly defined types of wool than the QDRs in the old National Wool Grading Standard (Section 5.1.2). Nevertheless, the concept that start (and reserve) prices for auctions should be established on the basis of predetermined and fixed QDRs reflects a reluctance to abandon this important feature of the old regulated market system. However, it must be remembered that at the time the Chinese Auction Standard was being drafted, important overseas wool auctions such as those in Australia were operating with reserve prices which were administratively determined on the basis of fixed differentials for the various types of wool.

It is not possible to compare directly the QDRs in the old National Wool Grading Standard (Table 5.5) with the QDRs in the Auction Standard (Table 8.3) owing to the different levels of specification of wool types in each standard. That is, it is almost impossible to relate the three fine wool types in the National Wool Grading Standard with the 72 fine wool types in the Auction Standard. Nevertheless, it is evident that the National Wool Grading Standard did not penalise the very short staple lengths as much as the Auction Standard does, nor did it downgrade the lower grades of improved wool as much. Once again, this illustrates that the pricing system implicit in the old National Wool Grading Standard was designed to favour (or discount less heavily) the low-quality wool produced by the majority of herders. It also reflects the fact that the Chinese wool auctions were expected to focus on better-quality domestic wool.

8.3 Analysis of Auction Data

Information was collected for two of the three auctions held in 1991, namely those in Xi'an and in Huhehot. The 1991 auctions were the first to follow the Standard outlined in Section 8.2 and the last before the deregulation of wool markets in 1992. Indeed with the inactivity in Chinese wool auctions since 1992, the auctions held in Xi'an and Huhehot in 1991 represent the most recent operational auctions. The information available and summarised in Attachments 8A and 8B at the end of this chapter includes the full auction catalogues supplemented by the start prices and the actual prices received. The analysis was primarily aimed at identifying the type of wool offered at the auction, at discovering how the auction operated (including price determination) and at obtaining an estimate of the actual price differentials among the grades generated by the auction.

8.3.1 Xi'an Auction

All the wool offered at this auction came from the XUAR and the sale was organised by the TIC in XUAR. The auction was moved from Urumqi to Xi'an because of the threat of civil unrest in the XUAR at the time the auction was intended to be held. Altogether 11,718 bales or almost 1 million kg of greasy wool were offered. The TIC acted as agent both for all sellers and for some buyers and also provided the auctioneer. The cost of participating in the auction was set at 1% of the price received. To have the organiser of the auction acting as an agent for both buyers and sellers had the potential to create some conflict of interest for the TIC and reflected the rudimentary nature of the Chinese wool auction system.

Table 8.4 indicates that only better-quality wool was sold through the Xi'an auction. Two-thirds of the wool was of 64 count (fineness between 21.5 and 23μm) with the remainder being 66 count (20.1 to 21.5µm). Much of the 64s wool was at the finer end of its micron range, since the average fineness was only 21.54µm (Table 8.4). The wool sold at the Xi'an auction was considerably finer than typical wool grown in the XUAR or in China as a whole. The average staple length of 69.2mm masks some variation among lots, with one-third of the wool having a staple length above 8cm. Relative to comparable Australian fine wool, the wool was short, with one-third of the wool having a staple length between 6 and 7cm. Less than 2% of the wool, however, had a staple length below 5cm. Clean yields, although low by Australian standards, were higher (by up to 10 percentage points) than typical XUAR wool.

Table 8.4 Wool Characteristics and Prices Paid at the Xi'an and Huhehot Auctions, 1991

	Xi'an	Huhehot
	- - - - - - mean (standard deviation) - - - - - -	
Fineness (µm)	21.54 (0.61)	19.99 (1.15)
Staple length (mm)	69.21 (11.15)	69.00 (11.05)
Clean-wool rate (%)	50.24 (5.81)	43.84 (4.57)
Plant debris (%)	1.53 (0.46)	1.08 (0.63)
Start price (¥/kg clean)	29.14 (2.51)	28.77 (4.18)
Actual price (¥/kg clean)	28.79 (3.02)	28.94 (4.27)
Lot size (no. of bales)	120.80 (75.2)	163.90 (134.3)

Wool supplied to the Xi'an auction came from 19 production units which were either divisional PCC farms in the XUAR or major breeding farms or studs in the AHB network (Table 8.5). As the detail in Attachment 8A demonstrates, many of the lots consisted of around 100 bales (8.7 tonne greasy wool), although single lots of 429 bales and 274 bales were supplied by the Ziniquan sheep-breeding farm (or stud) and by the Ba Zhou domestic animal farm respectively. (The average lot size was 120.8 bales or around 10.5 tonne on a greasy-wool basis.) Table 8.5 also reveals that the principal producer/supplier to the auction was the famous Gongnaisi sheep-breeding farm or stud which supplied 1,332 bales of slightly finer wool than the average of other wool supplied. The Gongnaisi wool was also longer and had a higher clean-wool percentage. (For details on Ziniquan, Gongnaisi and other major fine wool sheep studs in China, see Longworth and Williamson, 1995.) The greatest variation across the different producers was in staple length and, given the importance of staple length in determining the value of wool, this explained much of the variation in average prices received by the different producers.

The average (over all lots) actual price of ¥28.79 per kg (clean) differed little from the average start price of ¥29.14 per kg (clean), reflecting the dominant influence of the start prices on the prices actually paid. Indeed, within each grade and even for each lot, actual prices differed only marginally from the start prices. Table 8.6 provides a comparison between the QDRs in the Auction Standard with the start prices and the actual prices received. Price ratios implicit in the start prices (and

Table 8.5 Selected Auction Characteristics by Supplier, Xi'an Auction

Supplier*	Total bales	Average lot size	Real net weight	Clean	Fineness	Length	Vegetable matter	Actual price
	(no.)	(bales)	(kg greasy)	(%)	(µm)	(cm)	(%)	(¥/kg clean)
121 Reg 7 Div agr prod corps	378	126.0	34,983.9	37.47	21.64	6.63	2.03	27.20
124 Reg 7 Div agr prod corps	474	79.0	40,331.8	49.81	21.49	7.32	2.37	28.93
142 Reg 8 Div agr prod corps	537	134.3	44,495.8	41.08	22.15	6.68	2.20	29.95
147 Reg 8 Div agr prod corps	420	105.0	36,588.2	41.72	21.73	7.18	1.77	28.13
151 Reg 8 Div agr prod corps	783	195.8	62,817.0	49.51	21.69	7.60	1.51	29.94
170 Reg 9 Div agr prod corps	480	96.0	40,122.0	55.41	20.78	6.68	1.21	28.46
23 Reg 2 Div agr prod corps	656	131.2	62,804.0	48.62	22.22	6.55	1.26	25.88
84 Reg 5 Div agr prod corps	327	81.8	25,833.7	52.44	21.50	6.27	1.90	26.83
87 Reg 5 Div agr prod corps	1,001	125.1	83,006.7	51.54	21.79	6.51	1.63	27.88
Ba Zhou dom animal farm	274	274.0	22,810.1	51.26	21.72	9.71	1.27	32.00
Bo le qian jin dom animal farm	695	115.8	63,097.7	48.79	21.87	6.63	1.79	28.14
Breeding-farm dom animal res inst	823	102.9	72,850.9	48.76	21.82	6.86	1.68	30.08
Sheep-breeding farm Wenquan	746	124.3	68,498.5	51.42	21.09	5.88	1.47	25.50
Sheep-breeding farm Gongnaisi	1,332	102.5	132,876.7	56.24	21.42	7.71	1.22	30.41
Sheep-breeding farm 2nd Div	94	94.0	8,004.9	51.43	21.97	8.45	0.48	31.00
Sheep-breeding farm 9th Div	168	84.0	15,187.4	52.00	20.67	6.33	1.17	26.60
Sheep-breeding farm Baicheng	1,262	126.2	103,622.6	52.88	21.42	7.20	1.18	30.16
Sheep-breeding farm Tacheng	839	139.8	72,977.7	46.78	20.99	6.35	1.22	28.08
Ziniquan sheep-breeding farm	429	429.0	31,912.5	51.83	21.73	7.32	1.07	29.80

*Abbreviations used in the supplier names are as follows:
Reg—regiment; Div—division; agr—agricultural; prod—production; dom—domestic; res—research; inst—institute.

Table 8.6 Selected Price Ratios by Grade of Wool, Xi'an Auction

Grade	Number of bales	Quality differential ratio*	Price ratio implicit in start prices	Price ratio implicit in actual prices
66AI	1,156	118	112.4	113.5
66B1I	654	112	107.3	108.1
66B2I	1,704	108	104.3	102.3
66C1I	785	97	82.7	83.6
66C2II	86	78	na	76.7
64AI	2,597	113	110.2	112.8
64AII	94	109	109.9	111.6
64B1I	1,112	108	104.6	106.6
64B1II	468	104	103.5	105.3
64B2I	1,892	104	101.3	101.3
64B2II	242	100	100.0	100.0
64C1I	365	94	81.6	83.7
64C1II	390	90	81.1	81.9

na, Not available.
*As specified in the Quality Standard of Auctioned Wool (Appendix H).

consequently those in the actual prices, given the marginal difference between the actual and the start prices) are broadly in line with those defined by the QDR in the Standard, especially for the main lots offered at the sale. The major exception was the shorter, lower-quality grades (C1I and C2II) which suffered price discounts considerably more than those specified in the Standard. This reflected the large stocks of these lower-quality wools at the time of the auction and consequently the limited buyer demand for them. Conversely, the price premium for the finer 66s wool (20.1 to 21.5μm) over the 64s wool (21.6 to 23μm) was only half that specified by the QDR in the Standard. As the start-price differentials remain within 90% of the basic price implied by the QDRs, the variations from the QDRs suggest there was more renegotiation of the start price rather than any activation of the reserve price. Nonetheless, Table 8.1 indicates that a substantial proportion of the wool offered for sale at the auction was passed in. Almost all of this wool was sold subsequently following private negotiations.

Some simple regressions, reported in Table 8.7, were performed to see how the various wool and auction attributes influenced the actual prices received (either freely determined or through manipulation by the start and reserve prices). In equation 8.1 of Table 8.7, actual prices paid at the Xi'an auction were regressed against fineness, the natural logarithm of staple length, vegetable matter, clean-yield percentage, a dummy variable for quality grade (equal to 1 for Grade II and 0 for Grade I) and lot size. The results reveal that only staple length and quality grade have a significant influence on price. The natural logarithmic relationship between staple length and actual prices is similar to that implicit in the QDRs (Section 8.2.3). A decline from quality Grade I to quality Grade II however decreases prices by ¥1.44 per kg (clean) which is almost two-thirds more than that implied by the QDRs. This result is due to the much lower prices received by these lower-quality grades at the auction. Fineness does not show a significant effect at the 5% level of statistical significance, primarily because of the narrow range of fibre diameters exhibited by the wool offered for sale. Lot size appeared to have no effect on prices received. Clean-wool yield also had no impact, principally because prices are on a clean-wool basis.

The relationships between some of the wool attributes such as fineness or staple length and prices received may vary depending on the quality of the wool. Equation 8.2 tested for these interactions or, specifically, whether the relationships between fineness, staple length or vegetable matter and price differed depending on the quality grade. The results indicate significant relationships between price and all three interactions, suggesting, at least under the depressed conditions of the Chinese wool market in 1991, that improvements in staple length, fineness or vegetable matter are even more important for the lower-quality grade wools.

Table 8.7 Regression of Actual Prices Against Various Wool and Auction Attributes[a]

	Xi'an		Huhehot	
Equation number	(8.1)	(8.2)	(8.3)	(8.4)
Independent variables				
Constant	-27.82 (5.45)**	-26.15 (5.65)**	13.41 (4.29)**	25.96 (4.61)**
Fineness (μm)	-0.271 (0.235)	-0.153 (0.24)	-0.439 (0.192)**	-0.158 (0.18)
Length[b]	15.36 (0.969)**	14.29 (1.05)**	12.79 (1.98)**	4.11 (2.18)*
Vegetable matter %	-0.576 (0.351)	-0.716 (0.368)*	-0.559 (0.433)	-0.372 (0.472)
Clean yield %	-0.029 (0.033)	-0.016 (0.034)	0.039 (0.065)	0.026 (0.051)
Quality grade	-1.44 (0.451)**	-0.164 (16.81)	-2.06 (0.671)**	-16.8 (6.93)**
Lot size	-0.0001 (0.002)	-0.0013 (0.0018)	0.0017 (0.0014)	0.001 (0.001)
Fineness x quality grade		-1.57 (0.86)*		-0.55 (0.303)*
Length x quality grade		7.38 (2.71)**		13.26 (2.47)**
Vegetable matter % x quality grade		1.55 (0.847)*		-0.248 (0.654)
Adjusted R^2	0.82	0.83	0.87	0.93
Degrees of freedom	87	84	32	29

[a]The dependent variable in all equations is actual prices. Figures in parentheses are the standard errors. The symbol ** indicates a significant t-value at the 5% level, while * indicates a significant t-value at the 10% level.
[b]Length is the natural logarithm of staple length measured in centimetres.

8.3.2 Huhehot Auction

With around 9,000 bales and almost 800 tonne of greasy wool offered, the Huhehot auction was slightly smaller than the Xi'an auction. However, there were less than half the number of lots auctioned, as the average number of bales in each lot at Huhehot (164 bales) was considerably higher than at Xi'an.

Summary statistics for the Huhehot auction appear in Table 8.4. The wool offered for sale at the auction was among the better wool produced in the IMAR and consequently it would have been some of the best wool grown in China. The wool was particularly fine, with an average fibre diameter of just under 20µm which is below the lowest micron measurement for any lot offered at the Xi'an auction. Half of the wool was between 20 and 21.5µm, while more than 40% of the wool was less than 20µm. On the other hand, average staple length and the standard deviation of that length were remarkably similar between the two auctions, although examination of each lot revealed a much more even distribution of staple lengths for the Huhehot auction, with no wool offered with a staple length of less than 5cm. Clean-wool percentages, which averaged out at 43.8%, were considerably lower than for the Xi'an auction. The difference between the wools offered at the two auctions reflects the generally higher clean-wool yields in XUAR compared with IMAR due to climatic and production system differences between the two autonomous regions. The wool presented at the Huhehot auction was distributed roughly evenly between the quality Grades I and II.

There were two modes of supply to the auction. Around half the wool (350t) was purchased by the IMAR TIC from producers (i.e. State farms in the production base and from the Wushen County SMC) at the usual local prices and then regraded and sold via the auction to processors. Second, sales were made through agents on behalf of State farms in the production base. (As early as 1986, a wool production base for State farms had been established for wool auction purposes. In general, the SMCs have not been part of the production base, while not all wool-growing State farms are in the production base either.) The TIC charged a 3% commission on agent sales. Most of the testing of wool sold by agents was by the Textile Research Institute rather than the FIB which tested the wool purchased by the TIC. In accordance with the regulations outlined in Section 8.2.2, all future testing will be conducted by the FIB.

In general, there were fewer though more varied suppliers to the Huhehot auction compared with the Xi'an auction, although the suppliers were again confined to the one autonomous region, namely the IMAR (Table 8.8). The lot size also varied much more than at the Xi'an auction, ranging from 18 to 885 bales. One of the principal sources of wool for the Huhehot auction was Wushen County (Banner), home of the famous Eerduosi or Erdos wool, which produced over one-third of all the wool supplied to the auction. The superior quality of this wool is reflected in the much higher price received for it than for the wool of any other producer (over ¥1 per kg [clean] more than the next best supplier). The Erdos wool is popular because: the count is above 60s; special grade wool has an average staple length >8cm; and the conversion percentage from clean-scoured to top is about 80% (Chinese wool is usually in the range 73 to 78%, with 75% being normal). Mills in Shanghai, Nanjing, Shenyang, Lanzhou, Huhehot and other places have shown great interest in Wushen wool. Total wool production in Wushen County in 1991 was 2,250 tonne (greasy). The Wushen County SMC purchased 100% of this wool in 1991. In 1989, the Wushen SMC participated in the Beijing wool auction but has not participated directly in any auctions since that occasion (Section 6.3.1).

The average price received at the Huhehot auction was similar to that at Xi'an. However, reflecting the greater diversity in the wool offered, prices in Huhehot varied considerably more. As at the Xi'an auction, actual prices received were strongly influenced by the start prices. The link between the prices received, the start price and the QDRs for the different grades appears in Table 8.9. Once again, the start prices broadly follow the QDR for most grades. Some of the lower-quality, shorter-staple-length, lots have much lower start prices and actual prices than those implied in the QDR schedules, while the premiums implicit in the QDR for the finer, longer-staple wools were also not realised.

The link between actual prices and various wool attributes at the Huhehot auction are revealed in Equations 8.3 and 8.4 in Table 8.7. As in the Xi'an auction, staple length and quality grade both had the major impact on prices received. Staple length had a smaller impact on prices at the Huhehot auction compared with the Xi'an auction, while the quality grade had a larger effect. Vegetable matter, clean yield and lot size also did not have a significant impact at the Huhehot auction. A primary difference in the quality-price relationship between the two auctions is that wool fineness has an effect which is statistically significant at the Huhehot auction. The effect of a rise in price of ¥0.44 per kg (clean) for each 1μm fall in fineness is slightly less than that implied in the QDRs. Incorporation of the interaction terms in Equation 8.4 indicated that staple length and, to a lesser extent, wool fineness do have a different relationship with prices received, depending on the quality grades.

Table 8.8 Selected Auction Characteristics by Supplier, Huhehot Auction

Supplier	Total bales	Average lot size	Real weight	Clean	Fineness	Length	Vegetable matter	Actual price
	(no.)	(bales)	(kg greasy)	(%)	(μm)	(cm)	(%)	(¥/kg clean)
Ba meng agricultural management bureau	660	165.0	39,666	47.72	21.09	6.63	1.11	22.93
Baiyinxile animal farm	1323	441.0	99,270	49.82	20.67	7.40	1.12	29.67
Gadasu breeding farm for domestic animals	616	205.3	49,120	50.37	19.69	6.87	1.64	29.87
Hao ku sheep farm	361	180.5	28,880	44.94	20.18	6.25	2.08	
Heilung kiang lin farm	334	167.0	20,000	41.80	22.40	7.18	1.44	23.75
IMAR textile company	222	222.0	19,980	41.00	22.56	5.50	1.49	22.90
Keqi domestic animal products company	360	120.0	32,400	41.27	21.27	6.83	1.40	
Wushen Banner	3497	166.5	362,437	40.99	19.63	7.02	0.46	31.04
Xing an meng agricultural management bureau	594	84.9	52,261	48.38	20.58	6.85	1.76	25.40
Zhe meng gao lin tun sheep farm	178	59.3	14,240	42.23	19.52	7.37	1.60	29.80
Zhe meng mo li miao	187	93.5	18,700	41.95	19.42	6.40	1.50	28.00
Zhe meng mo li miao sheep farm	308	102.7	30,800	43.87	19.12	7.23	0.73	29.80
Zheng bai qi wool factory	375	375.0	30,000	39.58	19.56	5.58	2.80	24.90

8.4 Is There a Future for Chinese Wool Auctions?

Introducing auctions in the late 1980s represented a bold step by the Ministry of Agriculture and MOTI to further wool-marketing reforms. Following the first auction in Urumqi in 1987, an additional 10 auctions were conducted successfully up to the end of 1991. The two auctions in 1992 were failures and none were organised in 1993. Zhang and Niu (1994) raised some immediate reasons for the lack of interest in Chinese wool auctions since 1992. Brown (1995) elaborated on some of these reasons and suggested some more basic underlying problems with the conduct and organisation of the auctions. In general, the problems hindering the development of wool auctions in China can be grouped into those relating to the inconvenience of participating in the auctions, those associated with recent external factors making wool auctions a less attractive marketing channel, and the more fundamental problems with the conduct and organisation of the auctions.

Table 8.9   Selected Price Ratios by Grade of Wool, Huhehot Auction

Grade	Number of bales	Quality differential ratio*	Price ratio implicit in start prices	Price ratio implicit in actual prices
70AI	1,137	122	116.7	117.2
70B1I	340	117	112.6	112.9
70B1II	78	112	110.7	111.0
70B2I	1,242	112	113.1	113.9
70C1II	548	97	89.1	89.5
70C1III	109	94	88.3	88.3
66AI	678	118	116.0	116.4
66AII	181	113	113.7	115.3
66B1II	1,080	108	109.1	111.0
66B2I	267	108	112.5	112.8
66B2II	938	104	107.9	107.8
66C1II	1,067	94	86.2	87.0
64B1II	167	104	90.0	90.4
64B2II	120	100	100.0	100.0
64C1II	220	90	81.1	81.5
SORT I	841	86	75.9	75.6

*As specified in the Quality Standard of Auctioned Wool (Appendix H).

Several factors currently make it inconvenient for market participants to operate in the auctions. For suppliers, the rules relating to objective testing and clip preparation are difficult to meet, given the current poorly developed state of fibre inspection in China, as outlined in Chapter 6. For buyers, strict application of the rules regarding the registration of buyers may have deterred some from participating in the auctions since 1992. Similarly, although payment and delivery conditions have eased since the 1989 auctions in Beijing, they are stricter than for other wool-marketing channels in China and so have discouraged some buyers from participating.

The two key external events in 1992 which have since impacted on wool auctions were the rise in imports (Section 11.1.2) and the deregulation of wool markets throughout the pastoral region (Section 3.6). With imports highly restricted in the 1989 to 1991 period, mills turned to the auctions as a means of sourcing the better-quality domestic wool. However, since 1992 access to overseas supplies has greatly improved (Section 11.2.2). At the same time, mills have been able to deal directly with State farms and county SMCs in the areas producing the better-quality wool. Consequently, while auctions helped the east coast mills to identify where the best-quality wool came from, the added costs of participating in the auctions referred to above made the marketing channels which emerged from the deregulation of the Chinese wool market in 1992 more attractive.

Even if some of the immediate issues or problems raised by Zhang and Niu (1994) could be resolved, it is clear from the analysis in this chapter and from the discussion in Brown (1995) that there are other underlying obstacles to wool auctions becoming an integral part of Chinese wool marketing. Not only did they account for a minor proportion of the total Chinese wool clip in the years when they were held, but the price setting was highly contrived. Prices were loosely tied to QDRs implicit in the Auction Standard while an implicit reserve price was also in place. In this respect, auction prices were established by reference to price ratios in much the same way that official State prices were set for SMC purchases. The auctions, therefore, could be regarded as having been merely a vehicle for "official" sales of high-quality wool. It is unlikely that any more wool auctions will be held in China until the rules are amended to allow prices to be freely determined without reference to the QDRs in the Auction Standard.

The problems evident in developing the auctions are typical of many attempts at marketing reforms in China. Authorities are reluctant to forsake completely the traditional manipulation of prices to achieve non-marketing objectives. Furthermore, to have any significant impact on Chinese wool marketing, the auctions need to be extended to include much larger quantities of both fine and semi-fine wool. The wool offered at the auction has not been "typical" Chinese wool but rather the very best wool in China and wool which, as mentioned above, is more suited to marketing channels other than auctions in the longer term. If auctions are to become a major and more lasting contributor to wool-marketing reform in China, then they need to accommodate the types of wool typically grown in China.

It is possible that freer auctions may prove to be an important transitional means of improving Chinese wool marketing, given the remote and commercially unsophisticated nature of Chinese wool production. However, the ultimate role of an old-fashioned "public outcry" auction in wool marketing is open to debate not only in China but also in Australia and elsewhere. The possibility of sale by description and the use of modern electronic marketing systems suggests that in the longer term, even in China, the sale of wool by open auction may not have a future. Many of the problems which currently prevent sale by description also pose difficulties for the auction system, in particular problems with fibre measurement and objective testing. Resolving these problems and making auctions a truer reflection of exchange between buyers and sellers are needed if auctions are to develop into a major marketing channel in China. In the past, the nature of the auctions prevented them from revealing the true preferences of buyers and so identifying important attributes which State farm managers and ultimately individual herders should have been seeking to incorporate in the wool they supplied.

The simple regression analysis of information from the Xi'an and Huhehot auctions in 1991, however, did reveal some insights about quality premiums and discounts in the Chinese market. The percentage of vegetable matter in the wool did not appear to influence premiums greatly, although these percentages are low by world standards and the tolerance level in the auction standard is also set at relatively low levels. Staple length was shown to be a critical factor, with premiums being generated which encourage wool producers to aim for wool of 8cm or more. It is debatable, however, whether these premiums alone are sufficient incentive to stimulate wool growers to overcome the fibre-length problem which is a major constraint and needs to be resolved if domestic Chinese wool is to be a bigger part of Chinese worsted production. Nevertheless, the results of the analysis suggest that efforts to resolve the fibre-length problem are just as important (perhaps even more important) than attempting to increase fineness. The long-term strategy of aiming towards finer wools may be of little value unless priority is also given to resolving the fibre-length problem. To date, the auction system has met only partly the need to create an agribusiness environment which encourages improvement in the locally-grown wool. For instance, the key attribute of tensile strength is not measured or recorded in the auction catalogue. More complete market reforms such as developing freer and more comprehensive auctions offer scope to highlight the problems with domestic wool and create the necessary incentives to correct these deficiencies.

Attachment 8A Xi'an Auction Catalogue and Prices

Lot number	Supplier	Grade	Number of bales (no.)	Original net weight (kg)	Fineness (μm)	Spinning count	Staple length (cm)	Plant debris rate (%)	Real net weight (kg)	Clean-wool rate (%)	Clean wool (kg)	Start price (¥/kg clean)	Actual price (¥/kg clean)
2180,2181	a	64AI,66AI	60	5543	21.14	66	7.3	1.81	5543	40.2	2228.3	30	29.6
2182,2183	a	64BII,66BII	220	20435.5	21.74	64	7.25	2.05	20240	37.82	7650.9	29	29
2184,2185	a	64CII,66CII	98	9338.2	22.04	64	5.33	2.23	9200.9	34.38	3163.7	22.5	23
2125	b	66AI	108	8912	22.34	64	8.13	2.36	8796	47.6	4182.5	31.3	30.4
2122,2126	b	66BI,66BII	248	20821	21.88	64	7.22	1.66	20296.3	40.01	8120.5	29.4	29.5
2127	b	66CII	135	11791.5	21.96	64	5.44	2.19	11614.5	40.92	4719.5	23	
2128	b		46	3897	22.41	64	5.94	2.59	3789	35.78	1322.2		
2103	c	66AI	429	31912.5	21.73	64	7.32	1.07	31912.5	51.83	16540.2	29.6	29.8
2102	d	64AI	372	29376	21.03	66	8.01	1.25	29376	51.71	15190.3	31.9	30.45
2104	d	64BI	288	23742	21.86	64	6.01	1.39	23742	47.96	11386.7	28.7	27.5
2101	d	64AI	77	5848.5	21.86	64	8.04	1.33	5800	52.55	3047.9	31.3	31.3
2160	d	64AI	46	3852	22.02	64	8.36	2.07	3899	45.83	1782.8	31.3	30.5
2105	d	64BII	94	8122	22.07	64	6.64	1.71	8049.7	43.58	3508.1	28.5	
2060	e	66AI	189	17025	22.24	64	7.43	1.38	17025	50.34	8570.4	29.6	29.9
2061	e	64BI	110	10066	21.24	66	5.7	2.27	10066	48.23	4814.1	23.7	24.1
2062	e	66BI	187	17293	21.75	64	6.56	1.5	17111.4	49.88	8535.2	28.7	28.7
2064	e	66AI	48	4483	22.12	64	7.42	1.82	4405	53.07	2337.7	29.6	29.7
2063	e	64BII	67	6560	21.78	64	6.04	2.08	6440.6	47.66	3060.9	28.3	28.3
2111	f	66AI	122	10084	21.63	64	7.02	1.78	10002.7	53.28	5330.4	29.6	29.6
2112	f	64AI	97	7863.9	21.72	64	8.94	2.81	7759.3	56.83	4323.1	31.3	
2113	f	64AI	59	5313.3	21.87	64	8.67	2.7	5257.5	47.15	2423.7	31.3	
2114	f	66BI	71	6104.7	20.78	66	6.65	1.82	6037.5	51.86	3130	29.5	28.5
2115	f	66BI	62	5884.9	21.18	66	6.08	2.88	5825.5	41.53	2342.4	29.5	
2116	f	64BI	63	5486.6	21.77	64	6.58	2.22	5449.3	48.18	2607.5	28.7	28.7
2050	g	66AI	190	16482	21.64	64	7.06	1.16	16333.7	49.09	8018.2	29.6	29.6
2056	g	64AI	23	1998	20.48	66	7.26	1.18	1974.5	53.74	1061.1	30.4	29.5
2057	g	64BI	11	1003	21.8	64	6.01	1.29	981	45.81	449.4	28.7	

Attachment 8A Xi'an Auction *(continued)*

Lot number	Supplier	Grade	Number of bales (no.)	Original net weight (kg)	Fineness (μm)	Spinning count	Staple length (cm)	Plant debris rate (%)	Real net weight (kg)	Clean-wool rate (%)	Clean wool (kg)	Start price (¥/kg clean)	Actual price (¥/kg clean)
2051	g	66BI	240	20660	20.48	66	6.0	1.18	20457.5	47.12	9639.6	29.5	29.5
2052	g	66CI	127	11526	21.14	66	5.39	1.34	11410.7	41.6	4746.9	23.7	23.7
2055	g	66BI	248	22092	20.39	66	6.35	1.18	21820.3	43.34	9456.9	29.5	
2132	h	64AI	35	3182	21.82	64	8.01	1.75	3182	46.89	1492	31.3	30.5
2133,2136	h	64BI,64BII	70	6296	22.18	64	5.56	2.06	6296	37.21	2337.1	23	23.3
2130	h	66AI	137	11754	21.23	66	8.52	1.77	11540.1	44.88	5179.2	31.9	30.6
2131	h	66BI	178	15862	21.7	64	6.64	1.5	15570.1	37.9	5901.1	28.7	
2020	i	64AI	141	13191.5	22.07	64	8.0	1.35	13032.2	53.33	6950.6	31.3	31.3
2024	i		87	9299.5	22.28	64	4.91	1.69	9220.5	42.95	3960.2		
2023	i	66BI	36	3456.5	22.41	64	6.19	1.41	3399	48.57	1650.9	28.7	28.2
2021	i	66AI	137	12560.5	21.89	64	8.01	1.74	12399.7	51.47	6382.1	31.3	31.3
2022	i	70AI	31	2681	21.2	66	7.49	1.79	2653.5	48.44	1285.4	30.4	30
2025	i	66AI	146	12179.5	21.18	66	7.02	1.65	11976	49.48	5925.8	30.4	30
2026	i	66BI	185	15588	21.12	66	6.03	2.01	15432	45.78	7062.5	29.5	
2027	i	64AI	60	4738	22.38	64	7.25	1.83	4738	50.06	2371.8	29.6	29.7
2140	j	66AI	82	6872	20.36	66	8.04	0.86	6872	57.92	3980.3	31.9	31.9
2144	k	66BI	102	8519	20.44	66	6.68	1.16	8519	54.61	4652.2	29.5	29.5
2141	k	66AI	90	7504	20.81	66	7.44	1.25	7440	58.82	4376.3	30.4	30
2143	k	66AI	115	9563	21.04	66	8.01	1.37	9440.6	58.36	5509.5	31.9	31
2145	k	66BI	109	9373	20.9	66	6.23	1.14	9215.5	54.67	5038.1	29.5	29
2147	k	66CII	64	5544	20.72	66	5.02	1.14	5506.9	50.61	2787	22.5	22.8
2148	j		86	8407	20.98	66	4.61	1.48	8315.4	46.07	3830.9		21.3
2150	l	66BI	126	10128.7	21.79	64	6.41	1.42	10128.7	55.89	5660.9	28.7	
2151	l	66BI	200	15920.5	21.74	64	6.4	1.74	15920.5	54.09	8611.4	28.7	
2153	l	66CI	120	9698	22.09	64	5.32	2.02	9698	52.04	5043.9	23	23.4
2152	l	66BI	171	14460.7	21.72	64	6.18	1.59	14360.9	50.06	7189.1	28.7	
2155	l	66AI	64	5223.8	22.18	64	8.02	1.47	5223.8	57.9	3024.6	30.3	30.3
2154	l	66CI	120	10452.5	20.65	66	5.18	1.37	10452.5	45.63	4769.5	23.7	

Attachment 8A Xi'an Auction *(continued)*

Lot number	Supplier	Grade	Number of bales (no.)	Original net weight (kg)	Fineness (µm)	Spinning count	Staple length (cm)	Plant debris rate (%)	Real net weight (kg)	Clean-wool rate (%)	Clean wool (kg)	Start price (¥/kg clean)	Actual price (¥/kg clean)
2156	I	66AI	168	14359.6	22.18	64	8.01	1.47	14217.4	51.39	7306.3	30.3	30.8
2158	I	66BII	32	3004.9	21.96	64	6.54	1.93	3004.9	45.31	1361.5	28	27
2714	m	64AII	94	8073.5	21.97	64	8.45	0.48	8004.9	51.43	4116.9	31	31
2715	n	64BII	49	5387	22.26	64	6.81	1.21	5267.4	45.24	2382.9	28	28
2712	n	64AI	380	35625.5	22.25	64	8.04	1	35055.5	54.42	19077.2	31	30
2711	n	64BII	37	3344	22.55	64	5.72	1.24	3344	48.95	1636.9	23	23.4
2713	n		50	6527.5	22.08	64	5.73	1.71	6527.5	44.32	2892.9		21.3
2710	n	64BI	140	12730.5	21.97	64	6.46	1.15	12609.6	50.15	6323.7	28.3	26.7
2036	o	66BI	143	11398.5	21.03	66	6.39	1.07	11398.5	49.97	5695.8	29.5	29.5
2032	o	66AI	132	10394.6	21.74	64	8.0	1.03	10286.5	55.28	5686.4	31	31
2031	o	66AI	151	11719.5	20.91	66	8.02	1.14	11719.5	55.03	6449.2	31.5	31.5
2037	o	64BI	93	8454.8	21.65	64	6.5	1.54	8359.3	51.49	4304.2	28.5	28.5
2033	o	64AI	156	12507.8	22.12	64	8.01	1.36	12390.2	54.37	6736.6	31.2	31
2035	o	66BI	119	9099.5	20.38	66	6.3	1.18	9014.5	50.21	4526.2	29.5	29.5
2030	o	66AI	80	8029	21.03	66	7.4	0.8	7953.5	54.38	4325.1	30	30.4
2034	o	66BI	124	12901.6	21.03	66	6.63	1.24	12794.5	52.13	6669.8	29.5	29.5
2039	o	64AI	111	8527.4	22.3	64	8.03	1.08	8413.9	54.53	4587.9	31	32
2038	o	64BI	153	11387.9	22.02	64	6.71	1.34	11292.2	51.39	5803.1	28.5	28.7
2005	p	66AI	100	9978	21.67	64	8.05	0.92	9869.02	56.55	5581	31	32.1
2009	p	64AI	100	9263	22.18	64	8.59	0.76	9263	57.87	5360.5	31.3	32.8
2004	p	66AI	100	9279	21.66	64	8.02	1.33	9279	56.42	5235.2	31	32.1
2007	p	64AI	100	10958	20.92	66	8.39	1.71	10849.5	54.89	5955.3	31.5	31.8
2002	p	66AI	100	10642	20.11	66	8.73	1.47	10642	52.76	5614.7	31.5	32.4
2008	p	64AI	100	9703	21.84	64	8.03	1	9703	58.26	5652.9	31	32
2001	p	66AI	98	9696	22.42	64	8.26	1.15	9696	60.28	5844.7	31	32

Attachment 8A Xi'an Auction *(continued)*

Lot number	Supplier	Grade	Number of bales (no.)	Original net weight (kg)	Fineness (μm)	Spinning count	Staple length (cm)	Plant debris rate (%)	Real net weight (kg)	Clean-wool rate (%)	Clean wool (kg)	Start price (¥/kg clean)	Actual price (¥/kg clean)
2012	p	66BI	186	18503	21.66	64	6.37	1.03	18482	55.86	10324	28.5	28.3
2000	p	70AI	94	8866	20.66	66	7.77	1.3	8866	57.82	5126.3	30.4	31.3
2003	p	66AI	99	9243.5	20.71	66	8.53	0.92	9243.5	56.68	5239.2	31.5	32.4
2013	p	66CI	24	2294	20.68	66	6.28	1.4	2294	53.57	1228.9		23.7
2010	p		117	13673.5	21.82	64	5.19	1.84	13243	49.94	6613.6	16	22.8
2011	p	64AI	114	11574	22.16	64	8.03	1.04	11446.7	60.25	6896.6	31	31.6
2041,2042	q	64BI	101	8286	20.76	66	6.29	1.1	8166.8	55.88	4563.6	29	27
2043,2045	q	64CI,66CI	130	12915	20.88	66	5.46	1.54	12807.8	50.76	6501.2		22
2040	q	64AI	130	11614.6	21.1	66	7.08	1.3	11463.6	56.37	6462	30	29.5
2042,2044	q	64BI	111	10200.3	20.98	66	6.4	1.17	10200.3	55.4	5650.9	29.3	
2046,2047	q	66CI	234	21017.6	20.82	66	5.47	1.57	20860	52.78	11009.9	23	23.5
	q		40	5038.5	21.98	64	4.59	2.16	5000	37.31	1853.5		
2160	r	64BI	60	4940	21.87	64	6.64	2.04	4940	52.9	2610.3	28	28
2161	r	66BI	65	5235.5	20.44	66	6.07	1.62	5235.5	50.71	2654.9	29	
2162	r	64AI	74	5404	21.91	64	7.19	1.61	5353.2	57.1	3056.7	29	29
2163	r	66CI	128	10305	21.76	64	5.16	2.31	10305	49.06	5059.2	23	23.5
2190	s	64AI	274	22810.1	21.72	64	9.71	1.27	22810.1	51.26	11692.5	31	32

Key to "Supplier" a—121st Regiment 7 Division agricultural production corps; b—142nd Regiment 8 Division agricultural production corps; c—Ziniquan sheep-breeding farm; d—151st Regiment 8 Division agricultural production corps; e—Bo le qian jin domestic animal farm; f—124th Regiment 7 Division agricultural production corps; g—Sheep-breeding farm Tacheng; h—147th Regiment 8 Division agricultural production corps; i—Breeding farm domestic animal research institute; j—Sheep-breeding farm 9th Division; k—170th Regiment 9 Division agricultural production corps; l—87th Regiment 5 Division agricultural production corps; m—Sheep-breeding farm 2nd Division; n—23rd Regiment 2 Division agricultural production corps; o—Sheep-breeding farm Baicheng; p—Sheep-breeding farm Gongnaisi; q—Sheep-breeding farm Wenquan; r—84th Regiment 5 Division agricultural production corps; s—Ba Zhou domestic animal farm.

Attachment 8B Huhehot Auction Catalogue and Prices

Lot number	Supplier	Type	Grade	Number of bales (no.)	Wool weight (kg)	Net weight (kg)	Clean-wool rate (%)	Clean weight (kg)	Fineness (µm)	Staple length (cm)	Plant debris rate (%)	Place of despatch	Start price (¥/kg clean)	Actual price (¥/kg clean)
3701	ha	12	70AI	126	11081.5	10951.7	46.32	5072.8	18.62	8.16	0.28	da	33.1	33.2
3702	ha	15	70BI	159	14373	14245.8	41.92	5971.8	18.42	6.59	0.3	da	31.9	32
3705	ha	15	70BI	140	12641	12529	39.45	4942.7	18.57	6.59	0.33	da	31.9	32
3703,3723	ha		66CII	171	19012.5	18875.7		7012.9				da	23.5	24.2
3703	ha	28	66CII	51	4907.5	4866.7	35.29	1717.5	18.6	5.34	0.22	da		
3723	ha	28	66CII	120	14105	14009	37.8	5295.4	20.84	5.35	0.72	da		
3715	ha	12	66AI	216	22219.5	22046.7	42.69	9411.7	21.02	8.25	0.33	da	32.6	32.7
3716	ha	15	66BI	208	23271	23104.6	37.7	8710.4	21.12	6.93	0.36	da	31.6	31.7
3719	ha	15	66BI	267	30732.5	30518.9	42.13	12857.6	20.48	6.74	0.58	da	31.6	31.7
3713	ha	15	66BI	149	16196	16076.8	38.96	6263.5	19.2	6.82	0.49	da	31.6	31.7
3714	ha	12	66AI	109	11134	11046.8	45.78	5057.2	21.26	8.38	0.8	da	32.6	32.7
3718	ha	12	66AI	235	24988.5	24800.5	44.11	10939.5	21.42	8.31	0.36	da	32.6	32.7
3722	ha	15	66BI	275	28225	28005	40.7	11398	19.12	6.61	0.62	da	31.6	32.1
3724	ha	12	70BI	92	9755	9681.4	43.72	4232.7	18.93	7.9	0.31	da	31.9	32
3721	ha	12	70AI	238	22843.5	22653.1	41.09	9308.2	17.63	8.04	0.21	da	33.1	33.5
3725	ha	15	70BI	138	15618	15507.6	38.91	6034	18.12	6.61	0.13	da	31.9	32
3731	ha	12	70AI	99	9165.5	9086.3	46.78	4250.6	17.79	8.03	0.46	da	33.1	33.2
3732	ha	15	70BI	323	32022	31763.6	42.03	13350.2	18.7	6.95	0.45	da	31.9	32.4
3733	ha	28	66CII	169	19333.5	19198.3	35.18	6754	20.25	5.08	0.58	da	23.5	23.6
3717	ha	28	66CII	94	10918.5	10848	35.2	3818.5	21.1	5.01	0.73	da	23.5	23.6
3712	ha	12	66AI	118	12710	12621.5	47.96	6053.3	21.32	8.7	1	da	32.6	32.7
3602	hb	61	Sort I	108	6577.5	6491.1	47.21	3064.4		6.65	0.66	db	22.8	23.2
3604	hb		Sort I	280	17052	16828	54.48	9167.9		6.83	0.86	db	17.3	17.7
3601	hb	25	66BII	106	6455.4	6370.6	42.01	2676.3	21.09	6.49	1.4	db	28.8	28.9
3720	hc		60CII	222	20202	19980	41	8191.8	22.56	5.5	1.49	da	22.8	22.9
3401	hd	22	66AII	163	14883	14670	51.72	7587.3	21	9.16	1.21	dc	32.3	32.4
3402	hd	25	66BII	885	60515	59850	50.82	30415.8	20.46	7.21	1.2	dc	31.1	31.2
3403	hd	28	66CII	275	25025	24750	46.93	11615.2	20.54	5.84	0.96	dc	25.3	25.4
3431	he	28	66CII	375	30281.3	30000	39.58	11874	19.56	5.58	2.8	dd	24.8	24.9
3304	hf	28	64BII	81	6561	6480	45.87	2972.4	20.29	6.4	2.12	de		

Attachment 8B Huhehot Auction (continued)

Lot number	Supplier	Type	Grade	Number of bales (no.)	Wool weight (kg)	Net weight (kg)	Clean-wool rate (%)	Clean weight (kg)	Micron (µm)	Length (cm)	Plant debris rate (%)	Place of despatch	Start price (¥/kg clean)	Actual price (¥/kg clean)
3302	hf	25	66BII	280	22680	22400	44.01	9858.4	20.07	6.1	2.03	de	32.3	32.4
3221	hg	12	66AI	138	13938	13800	45.53	6283.1	18.76	8.4	0.78	df	32.3	31.6
3222	hg	15	66BI	128	12928	12800	44.12	5647.4	19.32	7.5	0.41	df	31.5	31.6
3223	hg	28	66CII	42	4242	4200	41.96	1762.8	19.28	5.8	0.99	df	25.3	25.4
3224	hh	28	66CII	78	7878	7800	43.85	3420.8	19.59	7.2	1.3	df	31.1	31.2
3225	hh	39	66CIII	109	11009	10900	40.05	4365.5	19.24	5.6	1.7	df	24.8	24.8
3231	hi	12	66AI	28	2268	2240	43.8	981.1	19.24	8.65	1.7	dg	32.3	32.4
3232	hi	15	66BI	120	9720	9600	43.4	4166.4	19.34	7.47	1.4	dg	31.5	31.6
3233	hi	28	66CII	30	2430	2400	39.5	948	19.98	6	1.7	dg	25.3	25.4
3201	hj	12	66AI	508	40986	40480	53.1	21494.9	19.98	8.5	1.6	dh	32.8	32.9
3202	hj	15	66BI	58	4698	4640.08	49.48	2295.9	19.34	6.6	1.6	dh	31.7	31.8
3203	hj	28	66CII	50	4050	4000	48.53	1941.2	19.76	5.5	1.73	dh	24.8	24.9
4096	hk		64B	167	10167	10000	40.84	4084	22.4	7.18	1.33	di	25.3	25.4
4097	hk			167	10167	10000	42.75	4275			1.55	di	22	22.1
3331	hl	25	66BII	120	10920	10800	39.94	4318.5	20.85	7.5	1.78	dj	30.2	
3332	hl	26	64BII	120	10920	10800	42.19	4556.5	21.68	6.2	0.99	dj	28.1	
3333	hl	61		120	10920	10800	41.67	4500.4		6.8	1.43	dj	22.8	
3112,3121	hm		66BII	124	10931.5	10819.9		5350.4				dk	30.6	
3112	hm	25	66BII	49	4886.5	4842.4	50.87	2463.3	20.83	6.9	1.7	dk		
3121	hm	25	66BII	75	8045	5977.5	48.3	2887.1	20.26	7.4	1.5	dk		
3113	hm		66CII	164	14603	14455.4		6687				dk	25.3	25.4
3113	hm	28	66CII	74	7433	7366.4	46.51	3426.1	20.6	5.8	2.1	dk		
3113	hm	28	66CII	90	7170	7089	46	3260.9	20.5	6.1	1.5	dk		
3111	hm	28	66AII	18	1726.8	1710.6	51.25	876.7	20.7	8.05	2	dk	31.6	

Key to "Place of despatch" da—Huhehot; db—Bamenglinhe; dc—Ximeng bai yin xile; dd—Ximeng zheng bai qi; de—Ke qi hao ku; df—Tong Liao; dg—Gao Lin tun; dh—Huanghua shan; di—Qiqihaer; dj—Keshiketeng qi; dk—Wu lan hot.

Key to "Supplier" ha—Wushen Banner; hb—Ba meng agriculture management bureau; hc—IMAR textile company; hd—Baiyinxile animal farm; he—Zheng bai qi wool factory; hf—Hao ku sheep farm; hg—Zhe meng mo li miao sheep farm; hh—Zhe meng mo li miao; hi—Zhe meng gao lin tun sheep farm; hj—Gadasu breeding farm for domestic animals; hk—Heilung kiang lin farm; hl—Keqi domestic animal products company; hm—Xing an meng agriculture management bureau.

Chapter 9

Local Scours or Local Scourges?

Few participants in the Chinese wool-marketing chain will fail to be affected by the major changes in wool-selling arrangements introduced in 1992. One group especially likely to be affected are the local, small-scale wool processors. Local processors are of interest for two reasons. First, by providing a value-adding industry related to animal husbandry, they are perceived to be important in the development of the strategically sensitive but geographically remote and industrially backward pastoral areas. Second, despite the small scale of the individual processors, the aggregate impact of a rapid expansion in the number of small plants that occurred in the 1980s adversely affected Chinese wool marketing and the wool industry in general.

The rapid development of additional local wool-processing units following the national fiscal reforms implemented in China in the early 1980s created several serious problems. Brown and Longworth (1992) highlight these difficulties. In this chapter, the expansion in local wool processing in China and its associated effects are reviewed briefly. The ways in which the most recent changes in wool-selling arrangements are likely to influence some of the matters raised by Brown and Longworth are then explored.

The fate of ill-conceived, small-scale scours in the remote pastoral areas may seem of little relevance to all but the localised regions they were intended to serve. However, the wool scours case lucidly illustrates some of the transitional problems that have arisen in China with the general economic reforms. In particular, it demonstrates how inappropriate incentives arising from uncoordinated and untailored national reforms can create major burdens for micro-level economic agents endeavouring to take advantage of these macro-level policy changes.

9.1 Development of Local Wool Processing

Following the formation of the People's Republic of China in 1949, wool processing was concentrated at large, fully integrated and centralised wool-processing mills. A series of national fiscal and related economic reforms beginning in 1980, however, encouraged local small-scale investment which in many areas included investment in wool processing.

The fiscal reforms replaced direct controls over the economy with more indirect fiscal and other measures aimed at favourably influencing the behaviour of economic agents at all levels in the economy (World Bank, 1988). In particular, local governments were assigned specific tax sources and were made responsible for meeting revenue targets and budgetary outcomes. The Central government retained the ability to redistribute wealth from richer to poorer regions by varying the specified tax sources and revenue targets assigned to local governments.

In some densely populated provinces on the east coast such as Jiangsu, the fiscal reforms led to an exponential increase in township enterprises. These new local enterprises were involved in many different low-technology manufacturing and processing activities including wool processing. For example, wool spindles in Jiangsu rose from 82,000 in 1980 to 644,700 in 1990, accounting for some 30% of all wool spindles in China at that time (see Section 2.3.2).

In the remote and sparsely populated pastoral areas where wool is grown, there was much less scope for fiscal-reform-driven local development. Limited industrial opportunities constrained the expansion of a revenue base required by the local governments to implement much-needed programs for social and economic reconstruction. Most of the local revenue base in the pastoral areas stems from animal production and animal husbandry-related activities (Longworth and Williamson, 1993). Consequently, the fiscal reforms in the pastoral areas led to much local investment in either upstream or downstream value-adding animal husbandry-associated activities such as feed processing, abattoirs and animal-fibre processing, especially wool processing.

For example, in the XUAR a significant number of extremely small, though fully-integrated, wool-processing mills were constructed at the local level in the 1980s. On average, these mills had a capacity of only 300 to 400 spindles and rarely exceeded 1,000 spindles. In many cases, these mills drew on local supplies of lower-quality crossbred wool and sold their wool products to local area markets. Despite their small individual size, their large numbers ensured they had in aggregate a non-trivial share of XUAR's wool-processing capacity. By the late 1980s, there were about 40 of these small county facilities in Xinjiang, although the depressed conditions of the early 1990s forced a number of them to close.

On the other hand, in the IMAR and Gansu, the fiscal reforms led to a rapid expansion in early-stage wool-processing factories, especially scouring plants, rather than to the construction of small fully-integrated mills. In the IMAR, there are only a few township-level scouring plants but there are more than 30 county scouring plants. In addition, each wool-producing prefecture has one or two prefecture-level scouring plants. There are also five large integrated wool-processing textile mills and one large topmaking factory in the IMAR, and three of these plants have scouring facilities.

Chifeng City Prefecture provides an interesting case study of how wool-processing facilities proliferated in the IMAR in the 1980s. During the first half of the decade, all wool-scouring facilities in Chifeng City Prefecture were attached to the large wool textile mills in Chifeng City, the capital of the prefecture (Fig. 9.1). Despite raw-wool production remaining relatively constant at about 10,000 tonne (Lin, 1990), total scouring capacity in the prefecture in the early 1990s has almost trebled since 1983 to around 30,000 tonne of raw wool. New scouring plants have been established in all five major wool-growing counties (Fig. 9.1). The result is a significant excess capacity of wool-scouring facilities in Chifeng City Prefecture. The excess scouring capacity has resulted not only from the number of counties wanting to scour wool but also because the minimum size of plants available (3,000 tonne of greasy wool throughput per annum) is more than double the annual raw-wool production in each of Aohan, Alukeerqin, Keshiketeng and Wongniute Counties and three times that in Balinyou County. In order to operate at or near full capacity, all the county scouring plants need to import significant quantities of raw wool from neighbouring counties or from even further afield.

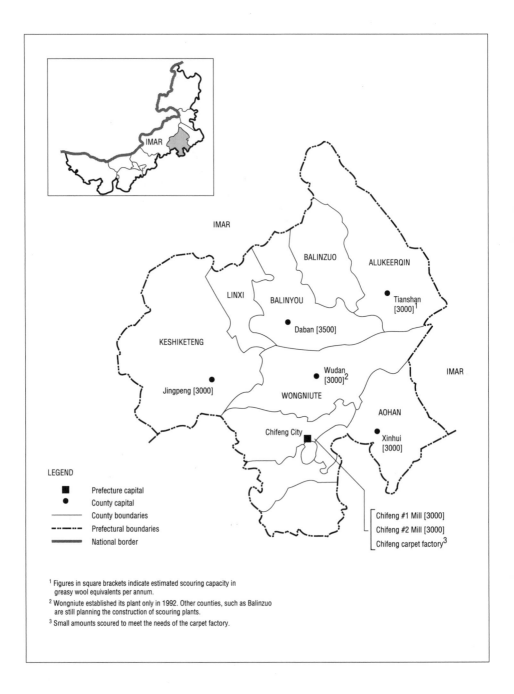

LEGEND

■ Prefecture capital
● County capital
—— County boundaries
— ·· — ·· Prefectural boundaries
▬▬▬▬ National border

[1] Figures in square brackets indicate estimated scouring capacity in greasy wool equivalents per annum.

[2] Wongniute established its plant only in 1992. Other counties, such as Balinzuo are still planning the construction of scouring plants.

[3] Small amounts scoured to meet the needs of the carpet factory.

Fig. 9.1 Location and Scouring Capacity of Wool-Processing Plants in Chifeng City Prefecture of IMAR

9.2 Adverse Effects of Rapid Expansion in Local Wool-Processing Capacity

Brown and Longworth (1992) highlighted the negative effects of the rapid development in county scouring facilities and their detailed findings will not be repeated in full. However, the essence of their arguments is outlined below because it illustrates the ease with which a well-intentioned national reform can have potentially disastrous effects at the local level.

Citing evidence from Chifeng City Prefecture, Brown and Longworth suggested that the national fiscal reforms led to local investments in scouring facilities which not only failed to meet the objectives of county governments but which also worsened the organisational efficiency of the Chinese wool-scouring industry as a whole and adversely affected the quality of scoured wool available to the Chinese textile industry.

The development of local wool-processing facilities was intended both to provide a source of fiscal revenue for the county governments and to generate new employment opportunities. However, many scouring plants in IMAR and Gansu Province incurred losses in 1990 exceeding ¥1.5 million, while most of the small fully-integrated wool mills in XUAR also incurred heavy losses. Thus local wool-processing plants have had to be subsidised by the county governments, exacerbating their budgetary problems rather than improving their fiscal situation. At the same time, employment effects have been rather modest as wool scouring is not a highly labour-intensive activity. County plants typically employ less than 100 people, many of whom are hired on a seasonal basis only.

The primary reason for the failure of wool-scouring plants to generate fiscal revenues for county governments was the high average scouring costs associated with the low levels of utilisation at these plants. These higher scouring costs (relative to larger, more centralised facilities operating at or near full capacity) more than offset any transport cost savings generated by scouring the wool closer to where it is grown.

In effect, the fiscal decentralisation policies of the Central government encouraged local governments to invest actively in particular local industries. However, the similar resource base of neighbouring communities led to parallel industrial development strategies in these neighbouring areas. The net effect was excess capacity in the total industry. Poorly developed information networks and market systems, along with narrow parochial perspectives, led local governments to pursue independently a course of industry development without regard to the effects of like developments in adjacent areas. Once the excess capacity is created, local governments attempt to protect local industries at all costs through "beggar thy neighbour" protectionist policies.

Another major concern about the development of wool-scouring facilities at the county level highlighted by Brown and Longworth was the poor quality of the scoured wool produced by these plants. The problem of low-quality scoured wool from county plants was found to exist throughout the pastoral region of China. Wool scouring is a demanding technical process and one which can lead to extensive and irreversible damage to the wool fibres if done incorrectly (Harmsworth and Day, 1990). Management of the scouring process is even more demanding in China than in other parts of the world owing to some characteristics of the Chinese wool industry. For instance, as pointed out in Section 6.6.2, the nature of the wool contaminants in the Chinese pastoral region means that scouring must be undertaken at higher temperatures, which compounds the risk of potential fibre damage. Furthermore, the poor quality of the locally manufactured detergents greatly increases the risks of scouring faults. Incorrect scouring procedures are almost inevitable at the new county plants, given the lack of experience of the managers at these plants. Staff at most of the new county plants have had no prior experience at scouring and often received less than one month of training at a large integrated prefectural mill before commencing operations at the county plant.

A further quality problem relates to wool grading. It is almost impossible to regrade wool after it has been scoured. The county-level SMCs which control and operate most of the local plants traditionally purchase the raw wool according to the imprecise old National Standard outlined in Chapter 4. They are supposed to re-sort it into industrial grades prior to scouring in accordance with State law (standards FG418 and FG423), which requires all wool to be graded according to the industrial-sort standard prior to scouring. However, the expertise of county-SMC staff to perform this regrading is limited (Section 6.3). Moreover, the final destination and required grade specification of the wool are often not known prior to scouring.

A priori, there are no good reasons why county plants could not devote more effort to training and quality control to improve the quality of their output. However, in the past the real problems have been: the traditional rigid price and marketing arrangements which have failed to provide incentives to increase quality control; the poorly developed information networks and coordination mechanisms which fail to highlight quality needs and deficiencies; the absence of capital mobility which prevents progressive plants from upgrading their capital base, including human capital; the restriction on foreign exchange and access to overseas imports, which prevents optimal levels of wool blending; and the lack of marketing reform to pressure those inefficient organisations enjoying State-conferred monopoly status to improve product quality.

Writing before the deregulation of the wool market in 1992, Brown and Longworth concluded that unless fiscal reforms were accompanied by appropriate market reforms, the fiscal reforms could not achieve the desired effect of more efficient resource allocation and greater regional development.

9.3 Likely Impact of New Selling Arrangements

Despite being set up in the belief that they would promote local development, by the beginning of the 1990s local scouring plants faced seemingly insurmountable problems and indeed represented considerable obstacles to further development in the pastoral region and in the Chinese wool industry. A pertinent question, therefore, is whether the significant changes in wool marketing in 1992 will have a favourable impact on the problems confronting local scouring plants. In keeping with the framework adopted by Brown and Longworth, the following sections examine this question in the context of industry organisational efficiency and wool quality.

9.3.1 Industry Organisational Efficiency

The boom times for the small local processors occurred during the period of the "wool wars" in the second half of the 1980s (Section 3.4). Demand for all wool, including poorly scoured wool from county facilities, was high. Furthermore, policies and arrangements regarding wool marketing had just been decentralised from the Central government to provincial, prefectural and even county governments. In this policy environment, county governments could grant their wool-handling agencies (usually the county SMCs) monopoly control over all wool produced within the county. Of course, in many instances the county governments had great difficulty enforcing the monopoly powers granted to the SMCs. It was the actions taken to protect these monopoly rights and the responses of those wishing to challenge the authority of the counties to grant such rights which created the chaotic marketing situation during the "wool wars".

The turbulent outcomes of the "wool wars", however, led to many of the provincial governments in the pastoral region reasserting their control over wool marketing in 1989 (Section 3.5). County SMCs could no longer direct all of their purchases to their own scours since they could be forced to send the raw wool to higher-level SMCs. On the other hand, as the market for wool collapsed in 1989 and 1990, the new centralised system provided a form of marketing "safety net" in that provincial authorities had some responsibility to acquire all wool, including locally scoured wool.

In the depressed wool market conditions of the early 1990s, the pastoral provinces experimented with relaxing the system. In particular, county SMCs were encouraged to find their own markets while the prefectural- and provincial-level mills were no longer obligated to take the scoured wool produced by county-level plants. Conversely, the county SMCs were not forced to sell their supplies to those mills.

Against this background, the re-opening of the wool market in 1992, outlined in Section 3.6 and elsewhere in this book, altered the situation for county scouring plants in two fundamental ways. First, at least in principle, herders and State farms could now sell their clips to whoever they wished and not necessarily to the local SMC or scouring plant. Second, the safety net for county processors which guaranteed the sale of their scoured wool was no longer in place.

Since 1992, county governments (and their SMCs) no longer have a legal monopoly over the purchasing of raw wool. Herders, State farms and other wool-growing entities can, in principle, sell direct to mills, or to a whole range of non-SMC buyers. They may also market their clips through wholesale markets and perhaps even wool auctions if such outlets are operating. Discussions with various herders and State farm officials immediately preceding the main wool-purchasing season in June 1992 revealed that these people welcomed the potential opportunities offered by multiple buyers. No "goodwill", "loyalty" or "tradition" was attached to the SMCs, with the producers intending to sell to whoever provided the best offer.

Despite the new competition, as discussed in Chapter 7, while the proportion of the national clip purchased by the SMCs has declined since 1992, the SMCs are still the overwhelmingly dominant buyers of raw wool. The SMCs have taken a much more pro-active approach since 1992 than they did after the initial opening of the wool market in 1985. Measures adopted include the introduction of mobile purchasing stations, providing better support services for herders, etc. (Section 7.4.3). Of course, the SMCs have two major advantages over other potential wool buyers: their enormous network of buying/servicing points and their close links with local and higher levels of government.

The importance of the latter should not be underestimated. As discussed in Sections 5.2.1 and 7.4.3, provincial governments may even be prepared to subsidise wool purchases by SMCs to enable them to remain viable social institutions in remote pastoral areas. Furthermore, county governments could be expected to adopt various measures to guarantee raw-wool supplies to their local plants. Some of these protection measures may be rather obvious, such as a compulsory system of registration or licensing for private buyers and special local taxes on their operations. But there are also many less formal ways in which local governments can support their SMCs. Nevertheless, despite the range of potential protectionist strategies available, in some cases it is virtually impossible for county governments to prevent herders selling their wool across the county border to neighbouring counties where buyers (including SMCs) may be actively seeking supplies.

To remain viable, many local SMC scouring plants will need to obtain a significant proportion of the wool supplies not only from within their own county but from neighbouring counties as well. As pointed out earlier, the standard minimum capacity of scours is around 3,000 tonne of greasy wool per annum, which typically exceeds county wool production. In order for these scours to operate at full capacity and achieve cost economies, they need to obtain wool from outside their county borders and so they must compete with neighbouring county plants for their raw materials.

Ideally, for quality-related reasons outlined in the following section, textile mills would prefer to purchase raw wool and scour it themselves or to have it scoured by specialist scourers or topmakers of their own choosing. In the more open wool market that is beginning to emerge, county scours have to compete not only with scours in neighbouring counties but also with direct mill purchases of raw wool. To survive in a fully deregulated market, scouring facilities will have to be located optimally in an industry organisational efficiency sense (i.e. be in the right location, have the right capacity and be serviced by relative cost-efficient assembly and distribution flows).

The evidence presented by Brown and Longworth suggested that economies of size in scouring are more important than transport costs and hence county sites are at a major disadvantage compared with large, more centralised facilities. In essence, most county capitals do not represent natural catchment centres for sufficient raw-wool supplies to maintain cost-efficient throughputs at scouring plants located at these county capitals. The resulting higher average scouring costs greatly exceed any transport cost savings achieved by scouring the wool closer to the producing areas. Thus, in a completely open market, only county scours which are servicing a reasonably large production area and which are geographically remote from other production areas and, therefore, from other county scours or prefectural/provincial mills are likely to survive without State assistance. In Chifeng City Prefecture, for example, perhaps only the scour in Alukeerqin could survive in a genuinely free market (see Fig. 9.1).

Despite the financial pressures county scours faced and the associated losses they were incurring, various counties were still planning to develop new scouring facilities in the early 1990s. A case in point was Wushen County in the famous Eerduosi Grasslands region of the IMAR which established its own scours in 1992. Although Wushen has some locational advantages as a site for a scouring plant, the local scouring of Wushen wool may prove to be a disaster. Some of the best fine wool in China is produced in Wushen and the surrounding counties. Unless the Wushen scouring plant is much better managed than other local scours in China, it will damage the high-quality raw wool and hence reduce its ultimate value to Wushen

County. In spite of these potential problems, Wushen County has opted for a twin set of scours with a capacity of around 5,000 tonne of greasy wool per annum or about twice the size of the standard minimum scours available and over double the annual wool production in the county. Thus Wushen County SMC not only is relying on obtaining supplies of wool from its own county but it also plans to purchase the best Eerduosi wool from the three neighbouring counties of Etuoke, Etuoke Qianqi and Yijinhuote.

Why do counties pursue such investments given the apparent losses they are likely to incur? In countries with a well-developed and comprehensive set of market and social institutions, the problems arising from conflicting local interests are generally of an ephemeral nature. Unfavourably-located plants or those producing a relatively poor-quality product face significant pressure to restructure in a well-developed market economy. However, in the PRC where only some reforms are in place and the marketing institutions are not well developed, the adverse consequences to local plants are more insidious and chronic in nature. It is in the transitional phase when reforms are being introduced in a piecemeal fashion that local development is most at risk. Even if the Central government perceives the broader consequences of the problem, it is unlikely to intervene for fear of reversing the otherwise beneficial effects of reform, or perhaps it is unable to take the necessary corrective action because the reforms generate a momentum of their own.

9.3.2 Wool Quality

Brown and Longworth identified the poor quality of the clean wool produced as one of the major problems associated with the development of county scouring facilities. A relevant question, therefore, is whether the change in wool-selling arrangements will create incentives which will encourage the county scouring plants to improve the quality of their product.

The more open marketing system will not necessarily lead to a set of prices reflecting the true value of scoured wool. The naive grading schemes described in Chapter 4 along with imperfect information and subtle forms of restriction on the exchange of clean-scoured wool between the producers and users of scoured wool will all inhibit the establishment of "fair" prices (see Chapter 5). Nevertheless, the changes in 1992 can be expected to raise the incentives for county scours to improve their operations. Because of the substantial impact of poor-quality scoured wool on later processing, such scoured wool will sell at a substantial discount if at all, especially during depressed market conditions. As already pointed out, county scours will be forced to compete with neighbouring scouring plants as well as with large mills scouring their own wool. In order to remain competitive in this environment, county scours need to establish themselves as a reliable source of quality scoured wool. This is especially so given the poor reputation they currently have among textile mill managers.

While the incentives to improve their operations have increased, can the county scouring plants respond to those incentives? Brown and Longworth emphasise that there are no fundamental reasons why county scouring plants cannot scour wool properly. However, there are a number of practical constraints which must be overcome, including inadequate training of staff in a technically demanding activity and insufficient fibre testing of scoured wool in China. In Chapter 6, it was argued that the more open marketing system, by itself, will not necessarily lead to better staff training or more appropriate fibre testing, at least not in the short run. For instance, in Section 6.3.1 it was pointed out that despite being aware of the need to regrade the wool properly according to industrial grades prior to scouring, managers in charge of the new Wushen scours had not taken steps to ensure that suitably trained wool-grading technicians were available.

The main concern of textile mill managers with county-scoured wool is that the wool is being either incorrectly graded prior to scouring or poorly scoured, or both. Better grading technicians at the county scours would undoubtedly improve the situation. However, there is a parallel need for the development of an independent testing service for scoured wool. Some plants do undertake their own testing but their expertise is limited and there is no system for checking or for obtaining feedback regarding the accuracy of their testing. It is argued in Chapter 6 that there needs to be a systematic effort to develop better and more widespread fibre-testing facilities. Such services are essential if county scours are to improve the quality of their product and satisfy the needs of textile mills.

Without these parallel reforms in relation to grading and testing, the opening of the wool market will exacerbate the quality problems county scours face. In terms of raw-wool input, the SMCs no longer have first offer of all wool in the county. The purchasing patterns for non-SMC buyers which are likely to emerge in the deregulated environment is purely conjectural. Nevertheless, it can be expected, for example, that large integrated mills, with their limited number of buyers, will purchase the better-quality wool. Wool sold through a reinvigorated auction system could also be expected to be of this type. Thus SMCs may be forced to purchase a larger proportion of lower-quality, less homogeneous wool. This would only compound the problems county scours face because it would be even more important and more difficult to re-sort such wool correctly prior to scouring. Poor-quality wool is also more difficult to scour properly.

More realistic price premiums for the different grades of wool generated by the open market may also lead to county scours purchasing a larger proportion of lower-quality, more heterogeneous raw material in order to maintain throughput while keeping costs down. However, this may not always be the case because in some instances, such as in Wushen County, the local SMCs may purchase the best wool in the belief that the final product they can obtain from their scours will be better and more marketable. In general, however, from the perspective of the whole Chinese wool industry, it may well be desirable to have the lower-quality wool scoured (poorly) at the smaller scours and the better-quality wool scoured (properly) at the large integrated mills. The recent opening up of new marketing channels and the removal of restrictions on the larger mills buying directly from State farms and others producing the better-quality wool are clearly moves in this direction.

9.4 Concluding Remarks

The re-opening of the wool market in many parts of the pastoral region from 1992 has placed further pressure on already depressed local wool-processing plants. Small inefficiently-located plants will have to close, or greatly improve the quality of their product in order to compete. In the short run, local governments will be under pressure to assist these enterprises to stay in business. In the longer term, there will need to be a major reduction in early-stage wool processing in pastoral areas. How the social costs of these adjustments are distributed could have a significant impact on the future of wool production in some of these areas.

The fiscal reforms of the early 1980s led to a surge of investment in local wool processing, but the agribusiness reforms in relation to wool marketing of the early 1990s have demonstrated how inefficient and wasteful most of this investment has become. Unfortunately, this is but another example of the high price China is paying for its "piecemeal approach" to economic reform.

The case of the wool scours also illustrates some of the opportunities and potential pitfalls that the market and economic reforms in China present to foreign firms. As the economic boom in some of the large urban areas on the east coast continues to attract a large proportion of the available investment capital, the Chinese government has called on foreign investors to consider meeting the investment shortfall in other regions such as the pastoral region. Certainly, foreign capital and skills could overcome some technical problems. However, local authorities may often seek external capital at almost any cost and for various reasons they may be poor judges of overall market developments. Thus foreign investors need to seek independent sources of information when considering investment at the local level.

Chapter 10

Problems Processing

The wool textile industry in China has experienced turbulent times since 1978. Traditionally, wool was processed in large integrated textile mills owned by the State and controlled by MOTI. During the early 1980s, many of the large State-owned mills located in the eastern provinces invested heavily in new imported equipment. At the same time, fiscal and other reforms provided both the funds and the incentives for local governments (at the township and county level) to invest in new industrial enterprises. Textile manufacturing was one such attractive "township enterprise", as these new ventures were termed, because it was a relatively low-technology undertaking and because cheap second-hand equipment was available from the larger State-owned mills which were re-equipping with imported machines. As emphasised elsewhere in this book, new township-enterprise textile mills, especially in Jiangsu Province but elsewhere as well, were responsible for the explosion in wool-processing capacity in the mid- to late-1980s. Since these mills needed access to raw wool, they played a major role in stimulating aggressive buying practices in the domestic wool market during the "wool wars" in China. Furthermore, the sudden surge in Chinese buying in Australia and other international wool markets in the 1986 to 1988 period, which helped to push raw-wool prices to record heights, was primarily the result of the sudden emergence of a new township-enterprise sector in the Chinese wool textile industry.

By the late 1980s, as discussed in Section 2.3.3, Zhang (1990a) and others could identify three distinct sectors in the Chinese wool textile industry, namely: the large relatively modern State-owned and MOTI-controlled mills located mainly in Shanghai, Nanjing, Tianjin and Beijing; the newly established township-enterprise sector in Jiangsu and other east coast provinces; and the up-country mills.

The collapse of the Chinese and international wool markets in 1989 created major problems for all three sectors of the Chinese wool textile industry, and most mills entered the 1990s in a parlous financial state. Despite an improvement since then, many of the underlying difficulties still remain. Some of these problems relate to Chinese trade policies and import/export arrangements and will be taken up in the next chapter. Another major source of difficulty for the Chinese mills, especially the up-country and other mills primarily processing domestic wool for the domestic market, has been the inefficiencies and lack of incentives for quality inherent in the marketing arrangements for domestic raw wool. These issues have been canvassed in considerable detail earlier in the book and only the main points are drawn together in this chapter to emphasise their impact on the wool textile industry. Three other major underlying constraints facing most Chinese wool processors in the 1990s are inadequate equipment, labour-force rigidities, and excess capacity. These three obstacles to profitability and progress are also examined.

The discussion in this chapter illustrates graphically the predicament of large State enterprises in China. General economic reforms have different impacts on various types of enterprises. For example, the township enterprises spawned by the general economic reforms operate under different and often more favourable labour arrangements, operating directives, State support systems and adjustment options compared with the older State enterprises. At the same time, a separate targeted reform agenda is being imposed on State enterprises to improve their efficiency. The combined impact of these general and specific reforms will force continuing structural change in industries such as wool processing, where State enterprises have traditionally had a dominant role. Among other things, this chapter and the next examine the ways in which State enterprises in the wool-processing industry are likely to be influenced by recent reforms, both of a general nature, such as labour-market reforms, and more specific reforms aimed at the wool textile industry.

10.1 State of Wool Processing in the Early 1990s

The Chinese wool textile industry at the start of the 1990s faced a major crisis. The crisis was not unique to wool processing but was endemic to the whole Chinese textile industry. Depressed domestic demand and substantial excess processing capacity meant that 37% of the textile mills were incurring losses in 1991 (Huang, 1991; Zhai, 1991). Profits had fallen by 70% in 1990 compared with the previous year and a further 24% during the first half of 1991. The lower profits were accompanied by increasing stockpiles of textile products, with stocks of cloth, silk products and woollen goods rising by 12.1%, 6.6% and 13.4% respectively in 1991 (Huang, 1991). Many of the township enterprises which had emerged during the 1980s either closed or were mothballed, while some of the State-owned enterprises required large government subsidies.

The economic downturn was more pronounced in wool processing than in the textile industry as a whole. For example, in the XUAR in 1990, only two wool-processing establishments were making profits, namely the Yili mill and the Tianshan Wool Product Company which also operated a joint-venture cashmere sweater factory. The small-scale integrated wool-processing plants operating at the county level in XUAR were especially hard hit and many were forced to close. The total loss incurred by the wool textile mills in XUAR in 1990 was estimated at ¥66 million. Losses in 1991 were reduced to ¥10 million, while most mills made a modest profit in 1992. In Gansu, one-third of the mills were incurring major losses in 1991.

The extent of the losses was apparently disguised somewhat by the accounting procedures used by the mills in relation to finished-product stocks. The stocks were valued according to administered prices rather than realistic sale prices. Thus any losses incurred on products stockpiled did not appear on the balance sheet until those stocks were sold. This process of delaying the accounting for perceived losses adversely affected the ability of mills to recover from the crisis as the demand for wool-based products improved from 1992 onwards.

In 1990, the Central government viewed with concern the financial circumstances of this important manufacturing sector. In the Eighth Five-Year Plan (1991 to 1995), MOTI adopted a hard line and sought to introduce a series of measures designed to improve mill efficiency. The problems of the wool textile industry were perceived to stem primarily from excess and antiquated capacity in the industry. Consequently, in the Eighth Five-Year Plan (8th FYP) there was a Central government edict that no decisions to create new capacity would be sanctioned and that the emphasis was to be on modernisation of existing processing capacity. At the time, the edict was to apply to all plants yet to be constructed, including those plants which had already received planning approval. In addition, MOTI threatened to impose punitive measures on plants with large stocks of finished products. Specifically, mills which had large stocks of unsold goods and which refused to cut production and adopt new technology would have their finance and energy supplies cut back (Zhai, 1991).

Implementation of the tough measures adopted in the 8th FYP has been influenced by developments in the Chinese textile industry since 1991 and by the institutional changes involving MOTI in 1993, described in Section 2.1.3. Nonetheless, the severity of the measures incorporated in the 8th FYP provides a clear indication of how seriously the Central government regarded the problems facing the wool textile industry at the beginning of the 1990s.

10.2 Domestic Marketing Arrangements and Wool Quality

Most Chinese domestic wool is of low quality by international standards. Typically, it includes a significant amount of short, tender and discoloured wool and it lacks overall uniformity of type. The general heterogeneity and, in particular, the non-uniformity of fibre fineness, length and tensile strength have major adverse effects on mill efficiency. As explained in Section 2.2, the growing of wool on a commercial basis has only a short history in China. There have been major investments aimed at improving the breeding and management of fine wool-producing sheep, and some remarkable progress has been achieved (Lehane, 1993; Longworth and Williamson, 1995). On the other hand, until the second half of the 1980s little effort was devoted to improving either the operational or the pricing efficiency of the marketing system (Longworth, 1993). Indeed, the crux of the quality problems with domestic wool has been the traditional marketing system which did not provide incentives for herders to supply better-quality wool. Further improvement in the marketing arrangements is an essential prerequisite for the continued success of genetic improvement programs and for the rapid adoption of better management practices.

Domestic wool-marketing arrangements and issues such as the (dis)incentives for quality implied in past arrangements have been canvassed at length in earlier chapters. This section highlights briefly how the past domestic marketing arrangements and recent changes in these arrangements may impact on mills using domestic wool.

To gain some appreciation of the extent of the quality problem facing Chinese mills, consider some characteristics of Chinese wool. Clean yields for fine wool average as little as 40% in Xinjiang and only 35% in the IMAR. The inherent environmental characteristics of these semi-arid pastoral areas clearly contribute to these low yields. However, management is also an issue, as reflected in the much higher yields (around 60%) achieved on some of the best State farms. Clean yields have also varied markedly over time, with deliberate adulteration of the wool in some years. For instance, clean yields fell dramatically (to less than 30%) during the "wool wars" in the second half of the 1980s.

Fibre length is also a serious problem. Both the weighted average length of fibres and the proportion of short fibres are of concern, although mills in the IMAR claimed that the former was the more serious problem. Fibre length was also claimed to be more of a production (management) problem than a processing (spinning technique) problem. Very little wool in the pastoral region is special grade fine wool (i.e. wool with a staple length greater than 8cm under the old National Wool Grading Standard). Indeed, in Gansu the average staple length is only 5.5cm. The analysis in Chapter 8 revealed that even the best-quality wool in China, namely that marketed through the auctions, is still characterised by low clean yields and short staple lengths.

For inland mills, the wool-marketing arrangements and the associated non-uniformity and adverse quality characteristics of Chinese wool have four main impacts. First, the heterogeneity of raw wool requires that the wool be re-sorted prior to first-stage processing. Re-sorting is labour-intensive and must be carried out by relatively skilled wool technicians. At the Ba Yi mill in Shihezi City in XUAR, for instance, one of the authors studied a two-stage sorting/grading operation which required a total of seven person-hours to process 100kg of raw wool. Most of the time was taken at the initial stage when the wool was removed from the bales and sorted into the required grades. The second stage of the grading process involved more experienced graders checking the initial sorting. As labour costs continue to rise in China, the pressure to avoid the re-sorting of the wool prior to processing will increase. The cost of this regrading in 1992 was about ¥0.15 per kg or 10% of processing costs.

Second, the non-uniformity in length and fineness reduces the conversion rates of raw wool to clean wool by 3 to 4%, scoured wool to tops by 2 to 3%, and tops to fabrics by around 2%. A top to noil ratio of 70:30 for Chinese wool compares adversely with a ratio of 80:20 for imported wool processed at the same mills.

Third, the heterogeneity creates scheduling problems for the mills. At the time of purchasing the wool, mills are not fully aware of its exact characteristics, and this creates problems in matching raw-wool supplies with intended mill outputs.

Fourth, low-quality raw wool inevitably leads to low-quality finished products despite mills making the best of the raw-wool supplies they purchase. For instance, in 1992 the manager of a XUAR mill which was exporting through a Tianjin mill estimated that a suit made from better-quality XUAR fabric would sell for ¥1,180 compared with ¥2,180 if it was made from a fabric imported from the United Kingdom.

The problems associated with low-quality wool increased throughout the 1980s. With greater freedom in relation to their production and management decisions under the household production responsibility system, herders are able to respond to the (dis)incentives implied in the marketing system. Furthermore, the reforms since 1978 have created a situation where much of the production and marketing risk is now met by individual households rather than collectivised by the State. Herders have responded to their more risky situations in the manner expected. They have diversified their activities and adopted more certain but less lucrative activities. As a result, in those areas where purebred fine wool sheep are not well adapted (because the environment is too harsh), there has been a reluctance to continue upgrading flocks by mating with better-quality fine wool rams. This has been one of the major reasons why the typical staple length of wool delivered to XUAR mills in the early 1980s was around 7cm but by the early 1990s it had dropped to only 6 to 6.5cm. It was also claimed by mill managers in XUAR that over the same period the proportion of fine wool Grade I which was 66 count (or better) had dropped from around 70% to less than 50%.

The regulated prices of the past meant that processors' margins were fixed (Section 5.1.4). Lower-quality products manufactured from the heterogeneous and low-quality domestic raw wool were not a real problem for processing mills. Under administered pricing, the main interest for the mills in marketing reforms and better-quality wool were on the cost side, namely better conversion rates, greater ease of scouring, etc. Nevertheless, processor returns could still be squeezed depending on how the administered prices and margins were set, especially when prices were used to achieve social objectives. For instance, discounts for lower-quality wool were relatively small irrespective of the weak derived demand for such wool, in part reflecting a welfare payment to the poorer herders who typically produce the lower-quality wool.

With the deregulation of the domestic retail market for wool-based products in the early 1990s, the quality of the fabric/yarn/garment became of major concern to the processors. Poor-quality products were difficult to sell in the comparatively sluggish retail market, even at much reduced prices. As already mentioned, MOTI threatened serious action against processors holding large stocks of finished products. Under these circumstances, the mills became more discerning purchasers of raw wool. The poorer-quality wool such as fine wool Grade II and especially improved fine wool Grades I and II were not in demand and, as pointed out in Section 7.3, the SMC was forced to finance substantial stockpiles of these wools.

In principle, the opening of the raw-wool market in 1992 should lead to more appropriate price differentials among grades. However, as discussed in Sections 5.2 and 7.3, local authorities still seek to influence raw-wool prices both directly and indirectly, and while the wool market is "open" it is far from "free" in some areas.

Price reforms alone, however, even when they are effective, may be insufficient to allow the clear transmission of price signals from consumers to producers and to enable mills to improve their efficiency. As argued in Chapters 4 and 6, appropriate grading systems and trusted inspection and testing procedures both need to be in place before prices can be expected to transmit appropriate signals through the long wool-marketing chain.

The old National Wool Grading Standard was clearly far too imprecise. The data discussed in Section 4.5 demonstrate the inappropriateness of the traditional grading system from the perspective of the mills. Mills could glean only limited *a priori* knowledge about the wool they bought on the basis of the grades defined in the old National Standard. This may not have been a major problem for the mills if the distribution of purchase grades across industrial grades remained relatively constant from lot to lot over time (and space). However, the evidence presented in Section 4.5 suggests that there were significant differences in the distribution across lots and across years. Under these circumstances, while the open market will tend to generate more appropriate grade-price differentials, if these prices still apply to broad, poorly specified grades, they will be only a limited improvement. The prices paid for these grades will not transmit to producers the true value that mills are prepared to pay for the various industrially relevant grades.

The new National Wool Grading Standard introduced at the end of 1993 represents only a modest improvement in relation to the grading of wool at the point of initial purchase. (See Section 4.3 for details.) On the other hand, the new National Wool Quality Control Regulation discussed in Section 6.4 and set out in full in Appendix G could, if widely implemented, put great pressure on market participants aiming to sell to mills to do so on the basis of proper sorting, grading and objective measurement.

The new trading regulations state that all transactions involving more than 2,000kg of raw wool must be conducted on a clean-scoured basis and that the wool must be properly graded according to the new National Standard. Furthermore, the wool must be objectively measured by an FIB or a designated official testing agency under the control of the FIB network. As the new regulations are gradually enforced, the SMCs and others wishing to sell significant quantities of wool to mills will have an incentive to regrade the wool prior to testing according to the Industrial Wool-Sorts Standard used by the mills. (See Section 4.4 for details.) As already mentioned, some State farms which previously sold through the auction system have now begun to sell their wool directly to mills on the basis of objectively measured industrial grades (Section 7.4.1).

One of the main problems identified in Chapter 4 in relation to the widespread adoption of more sophisticated and precise raw-wool grading is the lack of reliable grading, inspection and testing technicians. These issues are taken up in Chapter 6 where in Section 6.3 it is argued that mill managers generally have little confidence in the grading done by most grass-roots SMCs and on most State farms. Furthermore, mills also question the accuracy of measurements made by the official testing authorities. For example, Changji mill in the XUAR claimed that the FIB measurement of clean yield was too unreliable to use as a basis for pricing.

Under these circumstances, in the short run the new National Wool Grading Standard and the recently approved National Wool Quality Control Regulation will have only limited beneficial effects on the efficiency of the Chinese raw-wool market. More precise grading and trading on the basis of objective measurement both require substantial investments in the training of technicians and in testing equipment. Until these resources are available, only the best wool will be traded according to the Regulations. The bulk of the Chinese wool clip will continue to be marketed in ways which create major quality control problems for the mills.

10.3 Other Factors Affecting Mill Profitability

Many of the large State-owned textile mills in China pre-date the establishment of the PRC. Even some of the up-country mills, such as the larger plants in Lanzhou, the capital of Gansu Province in the centre of China, have a long history. The economic reforms during the 1980s highlighted the obsolescence and general poor quality of much of the equipment in these older mills. At the same time, as with all long-established State-owned enterprises, these mills have had to grapple with the burden of traditional Chinese labour-force policies. The rapid expansion in the total capacity of the Chinese wool textile industry also created an excess capacity problem for many mills.

10.3.1 Equipment

Various problems beset mills, especially those mills in the wool-producing regions, in their efforts to improve efficiency. Not the least of these is outdated, often locally made, machinery which has low technical efficiency and produces low-quality products. For instance, most Australian mills realise a top to noil ratio of 90 to 92%. However, even using imported wool, the equipment and techniques used by Chinese mills achieve a percentage of only 80 to 85%, with the very best mills achieving about 87%. Cheng (1991) cites statements indicating that 36% of all China's textile spindles are pre-1950 vintage, with another 23% being of low quality. The 8th FYP (1991–1995), as already stressed, explicitly acknowledged the obsolete equipment problem and it sought to stimulate mills to upgrade equipment rather than increase capacity. Incentives were put in place to encourage modernisation of old machinery. Following Zhai (1991, p.4), these included: tax incentives such as increased depreciation rates and shorter write-off periods; cash injections into the large State-owned mills in the coastal region to enable them to install new technology; and rules for the disposal of old equipment, specifically the disposal of obsolete equipment as scrap iron with penalties if it was sold to other mills.

A number of problems exist for mills aiming to achieve the modernisation targets set by the 8th FYP. First, many of the older mills had begun to modernise by importing new spindles, typically from Japan or West European countries, during the 6th and 7th FYP (1981–1990). In the early 1990s, some of these mills were still paying off old loans and could not readily increase their borrowings for further modernisation, given the depressed state of the domestic and international textile trade. Second, the losses incurred by many mills since the late 1980s, especially the least technologically advanced plants, restrict their ability both to obtain new loans for new machinery and to service these loans. Efforts to modernise have also been constrained by limited access to foreign exchange to purchase imported equipment. This has been an especially serious problem for the older up-country mills such as those in Lanzhou. These mills earn relatively little foreign exchange (since they produce mainly for the domestic market). Therefore, they have found it difficult to obtain the currency required to import up-to-date equipment.

One major difficulty facing Chinese mills in their efforts to improve product quality is that while their equipment is broadly suited to much of China's domestic wool it is not conducive to the production of high-quality worsted fabrics. Almost all combs in China are french combs, while spindles are of the ring type. French combs more profitably handle finer, shorter fibres (less than 5cm) while ring spinners are suited to medium wools. This has potentially serious implications for mills wanting to import wool in an effort to upgrade the quality of their final product. Attempts to upgrade product quality by purchases of high-quality imported wool may well be thwarted if the equipment available is not suited to processing that wool.

10.3.2 Labour Force

The introduction of new technologies and equipment as a means of improving mill efficiency requires commensurate changes in the structure of the workforce of the plant and the upgrading of skill levels. The long-established State-owned mills (which include many of the up-country mills) face a number of problems with their workforce which can be broadly grouped into those impacting directly on their productivity and those related to the financing of provision for their retired workers.

Within the first group, labour arrangements based on a cohort system, seniority and lifetime appointments pervade the State-owned enterprises. Allocating labour and determining wages according to seniority rather than merit undoubtedly affects the productivity of the mills. However, seniority is only one dimension of the problem. The favouring of particular groups over others restricts an efficient allocation of labour. And the problem has been carried across generations. Under the *ding ti* system, which first appeared in 1953, State enterprises had responsibility for recruiting the offspring of deceased or retired State employees, and this policy restricted the recruitment possibilities for these enterprises. Although the *ding ti* system was formally repealed in 1983, it persisted until 1986 when contract labour became the officially sanctioned method of recruiting new State workers. Davis (1988) and others have argued, however, that the contract system has tended only to institutionalise further the employment of cohort groups. Irrespective of developments since the mid-1980s, the legacy of seniority and *ding ti* still permeates many State enterprises and hampers their efficiency and labour productivity.

The large proportion of older workers, often with less formal education, may also hamper productivity at the long-established State-owned mills. Anecdotal evidence, however, is mixed. Wu (1993) in an examination of a limited number of rural textile enterprises found that both education and skill levels, along with other factors such as market agglomeration, management contracts, etc., were important in determining enterprise efficiency. The level of formal education was identified as an important factor for those enterprises which tend to recruit employees who are relatively less well educated. However, the skill levels of employees appeared to be related to experience and to on-the-job training. Therefore, up to some point, the longer the employees have worked for the mill the greater their skill level.

State-owned mills also face more rigid labour-relations arrangements regarding hiring and firing and other work practices than do collective enterprises (such as the township-enterprise mills). In general, the township enterprises have much greater flexibility as evidenced by their rapid and dramatic adjustments in response to changing market conditions. Much of the costs of these adjustments have been borne by their workforces. Conversely, adjustments by State enterprises to changing market conditions have been cushioned by subsidies from governments at various levels. Measures aimed at reforming the State enterprises seek to remove or limit these subsidies and so implicitly attack the rigid labour system existing in State enterprises. Overstaffing problems are a major manifestation of the rigid labour arrangements in the State enterprises. The contract system in which workers are given a less secure and less subsidised job is perceived by Chinese authorities as one way of addressing overstaffing. New technology that displaces labour can aggravate overstaffing. Given the costs or constraints associated with laying off workers, the State-owned mills have tended to expand their activities to soak up the workers displaced by new technologies.

The second group of labour problems for State-owned mills, especially up-country mills, concerns retired workers. The large State-owned mills typically have a significant number of retired workers on their payroll. For instance, the August 1st Worsted Mill in Shihezi in XUAR (a PCC mill with 14,184 spindles) had 600 retired workers on its payroll in 1992 or almost one-seventh of its active employees. The Shihezi mill was built in 1960 and many of the older State mills would have an even higher proportion of retired workers. Indeed, enterprises established prior to 1949 could have one retiree for every two or three workers.

For a developing country, China has had a long history of pensions, with State employees first being offered comprehensive coverage in 1951 (Davis-Friedmann, 1983). Initially financed as a supplementary wage bill, in 1953 the scheme was re-formulated as a national pension trust fund with mandatory contributions of 3% of the wage bill. However, the relative youthfulness of the workforce in the early 1950s and the exclusion of both farm workers and employees of urban collectives meant that there were few retirees. Throughout the 1950s and 1960s, the number of retirees gradually increased and the scheme was becoming too costly. Consequently, when the military took over State enterprises in 1969 at the height of the Cultural Revolution, it introduced a system where each enterprise paid for pensions out of its current wage bill.

The economic reforms since 1978 have had a marked impact on retirement benefits. Amendment of the Labour Insurance Regulations by the State Council in June 1978 (GUOFA No. 104) allowed pensions after only 10 years of continuous service rather than 20 years and promised jobs in the State sector to one child of each retiree (Davis, 1988). Retirement in the post-Mao era was no longer seen as socially unacceptable. Indeed, the government encouraged it as a means of removing the older workers and improving labour productivity and efficiency of the State enterprises. The incentives proved highly effective. In 1978, pensions took 2.8% of the wage bill whereas seven years later they represented 10.6% of the wage bill.

A key aspect of the 1978 and subsequent reforms was that while they initially encouraged retirement they did not represent a return to the original 1950s notion that pensions be collectively funded. In part, this reflected the policy dilemma being faced by Chinese officials. While wanting to encourage retirement to improve labour productivity, the demographics of a rapidly ageing population meant that this could be an expensive system to operate on a nationwide collective basis. Thus individual enterprises were required to carry the burden of the increase in retired workers.

As already stressed, the need to support retired workers places an especially heavy burden on the long-established State-owned textile mills because they have a relatively large number (and proportion) of ex-workers for whom they are responsible. Retirees receive not only substantial pensions but also free medical care and other benefits. The rates at which pensions were to be paid in XUAR in 1992 are set out in Table 10.1. The compulsory retirement age for technicians and cadres was 60 years for men and 50 for women. For ordinary workers, it was 60 years for men and 55 for women.

Table 10.1   Pension Entitlements for Employees of State-Owned Enterprises in XUAR, 1992

Retirement status	Years of service	Pension
		(% of full salary)
Employed before 1949		100
Retirement age	> 30 years	100
Retirement age	< 30 years	90
Retire early	> 30 years	100
Retire early	< 30 years	75

The demographic and social realities facing the Chinese government mean that it is loath to take on the financial burden of providing for the elderly at this stage. Filial care of the elderly has always been important in China both culturally and by regulation (Wen-hui, 1987). Apart from reducing the State's contribution to the elderly, filial care often has substantial private benefits, as the elderly provide child care and supplement household incomes with retirement pensions if they are former State-enterprise employees. Sometimes these pensions can exceed the current earnings of their descendants with whom they live. Changing social structures, such as the one-child family, may substantially alter the underlying nature of this filial care and may not be a panacea for Chinese authorities in dealing with the ageing population (Wen-hui, 1987). Some collectively funded pensions are in place but these tend to be limited to within localities and within industries. The economic reforms in the early 1980s, especially the ability to retain profits to improve productivity, mean that new enterprises with few pensioned staff strongly oppose moves to extend the scope of joint retirement funding.

The payment of retirement benefits out of current wages places unreasonable burdens on those (State-owned) enterprises with the least ability to meet them. Paying up to one-third of total wages to unproductive labour only compounds the inefficiencies of the older State enterprises and hinders their reform. For example, in the wool textile industry the combination of a growing proportion of retired workers, promotions according to seniority rather than merit, and life rather than contract employment, places the older mills at a distinct disadvantage *vis-a-vis* newer mills. For instance, compare the situation of the Gansu No. 1 mill (founded in 1940), which has around 17,000 spindles and employs 7,000 people, with the circumstances of the integrated worsted Xinjiang mill at Changji (founded in 1983), which has 11,300 spindles and employs only 3,000 people. Thus in responding to the micro-economic reforms confronting the Chinese wool textile industry, not all mills operate on a level playing field. China continues to grapple with the problems of labour-market reforms (see, for instance, Sun, 1992). The link between the older State-owned enterprises and the social welfare system complicates the nation's drive to greater efficiency in many key industrial sectors, including wool processing.

10.3.3 Capacity Utilisation

The growth in excess capacity in the Chinese wool textile industry in the late 1980s resulted from the rapid expansion in the industry (especially by the township enterprises) and the dramatic fall in imported wool supplies from 1989. The expansion problems in wool textile manufacturing were symptomatic of events occurring within the wider Chinese textile industry. Cheng (1991) states that the Chinese textile industry consisted of 1,045 mills with 26 million spindles in 1987 but just two years later in 1989 the industry had expanded to 2,088 mills and 35.6 million spindles.

The 8th FYP, as already mentioned, sought to cap this expansion by barring any investment in new capacity and allowing new spindles only when they replaced old spindles. Of course, the modernisation of existing equipment should improve spinning efficiency and other conversion factors, thereby increasing effective processing capacity.

Rationalisation in the wool-processing industry with a number of mills temporarily or permanently closed will relieve the capacity problem, as will the resumption of sizeable wool imports. However, in line with the arguments presented in Chapter 9, some of the underlying factors which led to the excess capacity problem still remain. Specifically, local authorities will continue to support the local value-added activities in the (often misguided) belief that they can generate employment and fiscal revenue for the local community. Frequently these decisions are made in ignorance of the state of the overall industry or even of events occurring in neighbouring communities. The problem is especially pronounced in localities such as the remote parts of the pastoral region where few other industrial investment opportunities are available. Thus chronic excess capacity could well continue unless other sources of fiscal revenue and employment opportunities are identified and local officials become better informed about developments occurring elsewhere in the industry.

10.4 Concluding Remarks

Major problems still beset wool-processing mills in China, especially those dependent primarily on local Chinese wool. Recent reforms to both domestic wool-marketing arrangements and wool-importing arrangements have the potential to improve both the quantity and the quality of raw-wool supplies to these mills and their efficiency in using these inputs. However, as in the case of the local wool scours and fiscal reforms discussed in Chapter 9, a comprehensive approach to policy reform is needed. That is, domestic and imported wool market reforms alone may not be adequate or even desirable if other reforms such as those relating to grading and testing are not also in place.

Furthermore, other underlying problems facing Chinese wool-processing mills and identified in this chapter, including outdated equipment and an inefficient workforce, need to be addressed. A lasting resolution in these areas will be dependent upon economy-wide reforms and policy adjustments in labour and financial markets.

Chapter 11

Trading Places

Wool imports are a key but volatile factor in the Chinese wool textile industry. While China has a long history of importing wool, the amount of raw and semi-processed wool flowing into the country expanded sharply after 1980. On a clean-wool equivalent basis, the quantity of wool imported exceeded domestic production by a healthy margin in all years after 1984 except for 1990. In fact, imports were almost double domestic production in terms of clean-wool equivalent in 1988, 1992 and 1993. Traditionally, imported wool was used primarily by the large, export-oriented, coastal mills. As discussed in Chapter 10, the sudden emergence of new township-enterpise mills in the second half of the 1980s greatly expanded the number of wool textile spindles available and hence the need for imports.

Recent changes in relation to Chinese wool-importing arrangements are of great interest since these reforms are likely to have effects not only on the domestic Chinese wool textile industry but also on international markets for wool and wool-based products. Chinese exports of wool-based yarns, garments and textiles now play a major role in international markets. With the constraints on the quantity and quality of domestically-produced wool, the growth in Chinese exports of wool-based products has been fuelled by wool imports. Indeed, China has become one of the major wool-importing nations.

The sudden emergence of China as a major importer of raw and semi-processed wool has been especially significant for Australia, the world's largest producer and exporter of wool. Prior to 1980, Australia shipped less than 10,000 tonne (greasy equivalent) of wool to China each year. As Fig. 11.1 demonstrates, these shipments were a negligible proportion of the total annual wool exports from Australia. However, Australian exports to China grew rapidly in the first half of the 1980s and by 1986/87 had reached more than 100,000 tonne (greasy equivalent). After a major slump which saw Australian wool exports to China drop to around 30,000 tonne (greasy equivalent) in 1989/90, the Sino-Australian trade in wool expanded once more to reach 180,000 tonne (greasy equivalent) or 22% of all Australian wool exports in 1993/94 (Fig. 11.1). In little more than a decade, China has changed from a relatively insignificant buyer of Australian wool to the Australian wool industry's single largest customer.

Wool imports and wool product exports have become important issues in China's attempts to rejoin the General Agreement on Tariffs and Trade (GATT) and to become a founding member of the proposed new World Trade Organisation (WTO). Imports are seen by some elements of the Chinese government as a threat to the livelihood of the sheep-raising minorities who live in the strategically sensitive northern and north western pastoral provinces of China. There are fears that unless some special dispensation for wool is negotiated before China enters the GATT (and the WTO), the domestic wool-growing industry will be disadvantaged severely and

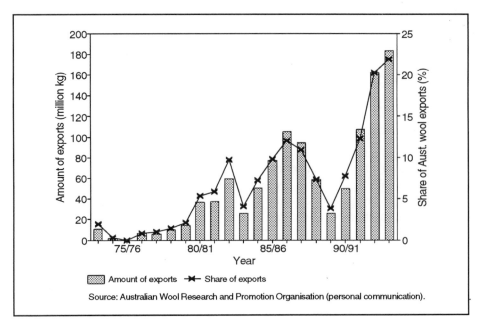

Fig. 11.1 Australian Wool Exports to China (greasy equivalent), 1973/74 to 1993/94

this will lead to political unrest in the pastoral region which occupies at least a third of the nation (Longworth and Williamson, 1993). In the past, Chinese wool product exports have been hindered by the contentious Multi-Fibre Arrangement by which international wool textile and garment trade has been traditionally regulated. If China is to enjoy the removal, albeit gradual, of the Multi-Fibre Arrangement negotiated in the GATT Uruguay Round, then it needs to attain full membership of the GATT/WTO. Wool-trading issues, therefore, are of major significance for China as it seeks to become a full member of the international trading community.

This chapter is primarily concerned with outlining the many recent changes in the institutions, policies and administrative arrangements influencing the importation of wool. The first section provides some important background information but the remainder of the chapter is devoted to an analysis of how the recent reforms will impact on the domestic wool textile industry. Traditionally, Chinese wool-importing arangements have been heavily biased towards mills well positioned to re-export their wool-based products. In some respects, the changes since the late 1980s have re-enforced this bias. However, the recent reforms also have the potential to expand greatly the demand for imported wool as up-country and other domestic-market-oriented mills, which have traditionally had only limited rights to the use of imported wool, gain freer access to overseas supplies.

11.1 Background

The recent institutional, trade policy and administrative reforms of relevance to the Chinese wool trade have been both the cause and the result of the dramatic changes in the flow of wool imports into the country. As background to the discussion of these reforms in Section 11.2, some statistical details concerning Chinese wool imports in recent years are provided in this section. First, however, some comments are made about the available data.

11.1.1 Limitations of Chinese Wool Trade Statistics

It is difficult to obtain reliable statistics on the quantity of wool imported into a country such as China. Indeed, a great deal of caution needs to be exercised in interpreting wool trade statistics, especially those emanating from Chinese sources. Traditionally, there have been two distinctly different sources of official Chinese trade data, namely the Chinese Customs and MOFERT/MOFTEC. In the case of wool, there have been major differences in the data provided by these two agencies.

Collecting statistics on wool imports is especially complicated because wool is traded in various forms such as greasy (or raw) wool, scoured and/or carbonised wool, and tops. Conversion factors are used to express the quantities traded in the various forms in terms of a common numeraire, usually "clean-wool equivalent" or "greasy-wool equivalent". It is important, therefore, when using wool trade data both to take note of which numeraire is being used and to be aware of the conversion factors applied by the various data-collecting agencies. To complicate the situation further, different agencies publish wool trade statistics for different time periods. For example, Fig. 11.1 is derived from greasy-wool equivalent data published by the Australian Wool Research and Promotion Organisation (AWRAP) (formerly the Australian Wool Corporation and from September 1994 merged with the International Wool Secretariat (IWS)) on a July to June year basis. On the other hand, the Chinese Customs statistics used to construct Fig. 11.2 are collected in clean-wool equivalents on a calendar-year basis.

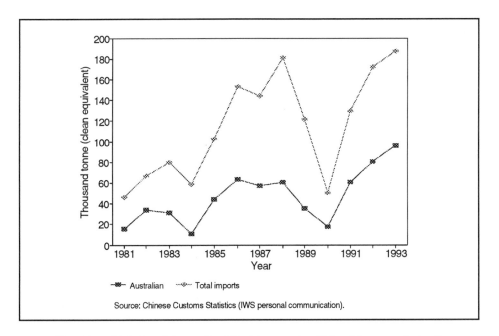

Fig. 11.2 Global and Australian Wool Imports by China (clean equivalent), 1981 to 1993

While the data for total imports shown in Fig. 11.2 are likely to be reasonably accurate, Chinese wool import statistics disaggregated by micron range, by wool form, or by country of origin are much less reliable. For instance, the whole of each parcel (or lot) of wool will be assigned to a particular mean fibre diameter (micron) range according to the fineness of the dominant line even if that line accounts for less than half the lot. As regards separate import statistics for the various forms of wool, scoured wool and carbonised wool are frequently lumped together by Chinese import data-collection agencies. Import statistics classified by country of origin are especially problematic.

The data in Fig. 11.1, for example, refer to exports shipped directly from Australia to China. On the other hand, the relevant graph in Fig. 11.2 indicates the amount of raw and semi-processed wool imported into China and declared to be of Australian origin irrespective of whether it was shipped directly or not. After making the necessary adjustments for different numeraires, time periods and conversion factors, the latter figures can be expected to exceed the former. Furthermore, topmakers commonly blend Australian wool with wool from other sources but still label their product as "Australian Top" to command a premium. To the extent that these tops are imported by China, the Chinese Customs data plotted in Fig. 11.2 for Australian wool imports (which include tops) may overstate the true situation.

It is also difficult to draw accurate comparisons between the quantity of domestic wool grown and the quantity of wool imported. Even for national totals, it is only possible to get rough approximations in clean-wool equivalent terms. For example, the 1991 Chinese domestic clip was about 240,000 tonne of greasy wool. On a clean-wool basis, this was equivalent to around 100,000 tonne, which can be compared with wool imports of approximately 130,000 tonne (clean-wool equivalent) in that year.

11.1.2 Recent Trends in Chinese Wool Imports

While the limitations of the available data must be kept in mind, some remarkable changes have occurred in Chinese wool imports since 1980. Reference has already been made to the dramatic expansion in total imports of wool during the 1980s which reached a peak of around 180,000 tonne (clean equivalent) in 1988. Not only was this quantity about four times the level of imports in 1981 but, as mentioned earlier, it was also about double domestic production in that year. Economic and political conditions in China in 1989 and 1990 "killed" the previously overheated domestic market for consumer durables such as wool-based goods. At the same time, a major glut of wool and wool-based products emerged on international markets, and overseas markets for Chinese wool textiles contracted sharply. As a result, wool imports into China shrank to less than 50,000 tonne in 1990, which was about the level of imports before the dramatic expansion of the 1980s. But imports bounced back in 1991 to more than double the level of the previous year and continued to grow strongly so that by 1993 imports exceeded the previous record levels achieved in 1988 (Fig. 11.2).

In recent years, there have been significant shifts in the mix of wool imports in terms of both the form in which wool is imported and the country of origin. The available statistics classify imports into three forms: greasy/raw wool, scoured (and/or carbonised), and tops. As Fig. 11.3 demonstrates, the proportion of imports represented by greasy wool dropped from 30% to less than 10% during the 1989/90 slump but recovered to around 30% again in 1991. After 1991, the proportion of wool

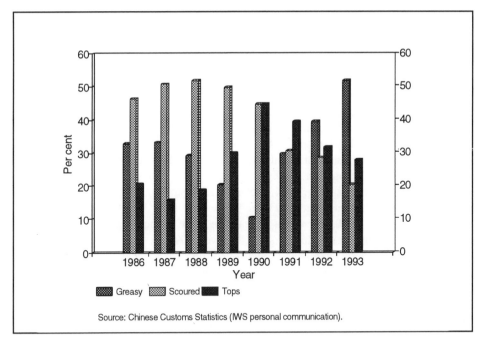

Source: Chinese Customs Statistics (IWS personal communication).

Fig. 11.3 Share of Chinese Wool Imports by Form of Wool, 1986 to 1993

imported in the greasy form continued to increase, exceeding 50% in 1993. Given that total imports dropped from 180,000 tonne in 1988 to 50,000 tonne in 1990, the reduction in the proportion of wool imported in the greasy form in 1989/90 is remarkable. Clearly, greasy-wool imports were the segment of the trade most severely affected by the collapse of the Chinese market in these years. In contrast to greasy wool, the proportion of tops imported rose sharply during the 1989/90 slump. In fact, tops more than doubled their share of imports from 20% to 45%. Since 1990, however, both tops and scoured wool have lost market share to greasy wool (Fig. 11.3).

The reasons for the sudden shifts in preferences in relation to the form of wool imported into China are likely to have been both administrative and commercial. As will be discussed in the next section, the Central government recentralised control over wool imports in 1989 and this move was motivated, in part, by a desire to prevent Chinese buyers competing "irrationally" with each other in international auction markets for greasy wool. At the same time, there were commercial incentives for mills to buy tops rather than raw or scoured wool to shorten the processing period and hence price risks. Another factor favouring the importation of tops may have been the unsatisfactory experiences some mills had with raw wool imports (Section 11.4) and the problems with early-stage wool processing in China (Chapters 9 and 10). The swing back to importing greasy wool after 1990 probably reflects the easing of centralised administrative controls and growing confidence in the commercial future of the Chinese wool textile industry.

In relation to the country of origin, Australia has traditionally been the dominant supplier of the relatively finer apparel wool (i.e. wool with a mean fibre diameter ≤25µm), while New Zealand, Uruguay and Argentina competed for the stronger end of the apparel wool market (from 25 to 35μm mean fibre diameter) and for the carpet wool trade (>35μm). At the end of the 1970s, Australia and New Zealand together provided almost all the non-carpet wool imported by China. However, New Zealand's share of this segment of the Chinese market was steadily eroded by Uruguay and Argentina during the 1980s. Much of the South American wool was used by the new township-enterprise mills for products destined for the domestic Chinese market such as knitwear and hand-knitting yarn. Since domestic demand for these products contracted noticeably as a result of economic and political events in China in 1989, and as it is these kinds of products for which the bulk of the domestic clip is best suited, the Chinese demand for South American wool dropped sharply from 48,000 tonne in 1988 to 9,000 tonne in 1990.

In aggregate terms, as Fig. 11.4 shows, the Australian share of Chinese wool imports declined from just over 40% in 1986 to a little less than 30% in 1989. However, during the 1990s, there was a steady increase in the Australian market share of total Chinese imports so that it stood at over 50% in 1993. But imports from Australia represent a much higher proportion of the finer apparel wool than of all wool. For example, the Australian statistics for 1993/94 presented in Fig. 11.5 demonstrate that more than 90% of Australian exports to China in that year had a mean fibre diameter ≤27μm and that this has been the case in all but two years since 1982/83.

While the proportion of Australian wool exports to China which had a mean fibre diameter ≤27μm has remained relatively constant at around 90%, there has been a marked change in the distribution of this wool by fineness as the volume of the Sino-Australian trade has expanded. Specifically, almost half the Australian wool exported to China in 1982/83 and 1983/84 was <20μm. As the amount of Australian wool imported by China increased during the 1980s, the proportion of this very fine wool declined rapidly to around 5% in 1987/88. Fig. 11.5 shows that since 1985/86, wool with a mean fibre diameter in the range 20–23μm has made up two-thirds or more of all imports. At the same time the proportion of 24–27μm wool increased from around 5% in 1985/86 to a peak of over 20% in 1990/91 but declined to 10% in 1993/94. The combined proportion of 20–23μm and 24–27μm wool in Australian exports to China has remained remarkably stable at around 90% since the mid-1980s, but there appears to be considerable year to year switching between these categories (Fig. 11.5). This last observation implies that a large number of Chinese end-users of Australian wool are indifferent as to whether the wool they purchase is at the top end of the 20–23μm range or the bottom end of the 24–27μm category.

Not only is Australia the dominant source of apparel wool for China, but also China has emerged as the single most important destination for Australian wool exports (Fig. 11.1). Sino-Australian trade in wool now plays a major role in the international wool market and represents the single most important trade link between these two countries. The slump in Chinese demand for Australian wool in 1989 and 1990 placed considerable pressure on the ill-fated Australian reserve price and buffer-stock scheme for wool. (For a description of the demise of the reserve price scheme, see Gerritsen, 1992, pp.103–105.)

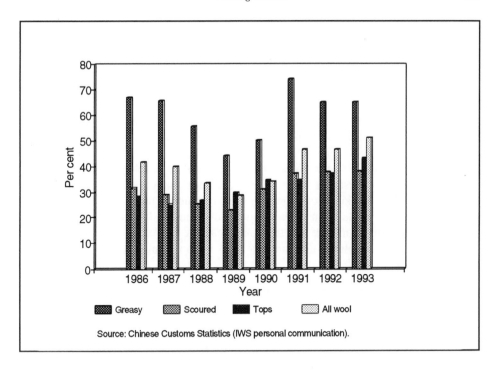

Fig. 11.4 Australian Share of Chinese Wool Imports by Form of Wool, 1986 to 1993

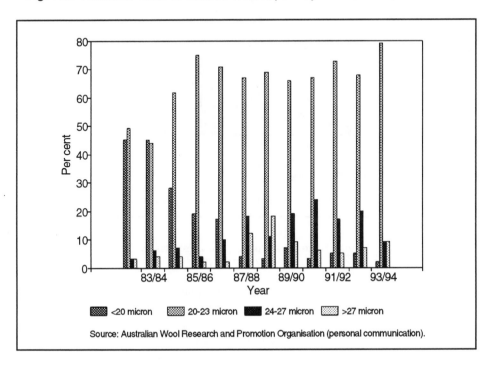

Fig. 11.5 Proportion by Micron Range of Australian Wool Exports to China, 1982/83 to 1993/94

Since Australia is the major supplier of Chinese greasy-wool imports (Fig. 11.4), and since it was greasy-wool imports which declined the most in 1989 and 1990 (Fig. 11.3), the Chinese share of Australian greasy-wool exports dropped from 10% to around 2% between 1986/87 and 1989/90 (Fig. 11.6). Despite this remarkable slump in the late 1980s, in terms of both physical quantities and values, annual exports of greasy wool from Australia to China have dominated the wool trade between the two countries in most years. However, on a market-share basis, a much higher proportion of carbonised wool and wool tops produced in Australia goes to China than is the case with greasy wool (Fig. 11.6). The expansion in Chinese imports of semi-processed wool since 1980 has played a major role in the development of early-stage wool processing in Australia. Given the difficulties many Chinese mills have in handling the particular kinds of vegetable fault commonly found in Australian wool (Section 11.4), this would seem to be a logical development. Indeed, as the up-country and other mills which produce for the mass domestic market in China gain greater access to overseas suppliers, they are likely to import semi-processed raw material rather than greasy wool.

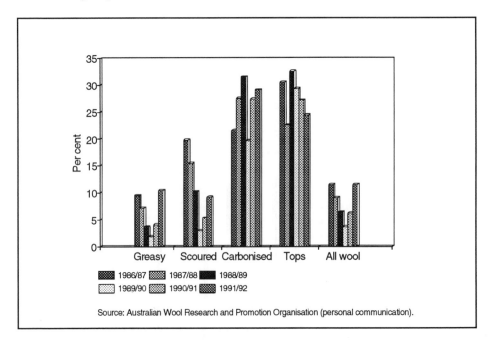

Source: Australian Wool Research and Promotion Organisation (personal communication).

Fig. 11.6 Chinese Share of Australian Wool Exports by Form of Wool,
1986/87 to 1991/92

In principle, imported wool competes with domestic wool. But wool is an extremely heterogeneous raw material. It is completely inappropriate to compare large aggregates such as total Chinese production with total imports and draw conclusions about the extent to which wool from these two sources competes. For reasons discussed in Chapter 2 and elsewhere in this book, Chinese wool is usually of poor quality compared with imported wool. Furthermore, China produces relatively little homogeneous wool of $\leq 25\mu m$ in mean fibre diameter. For example, Longworth and

Williamson (1993, p.329) argue that in 1991, while the total amount of wool produced of this type was about 46,000 tonne (clean), only about one-quarter of this wool would have been homogeneous enough to be considered comparable with the great bulk of the Australian clip of around 600,000 tonne (clean) in that year. In practice, therefore, domestic wool and imported wool tend to find different end-uses and are more complementary than competitive.

While it is difficult to obtain reliable data, it is widely considered that around 75% of imported wool is processed for the domestic market and the remaining 25% is for re-export. Since most of the Australian wool imported by China is ≤25μm in mean fibre diameter and is at the better end of the quality scale, it is highly likely that much more than 25% of Australian wool is re-exported. The proportion re-exported also depends upon the type of products being produced and the location of the mill. As mentioned earlier, in the 1980s the township-enterprise mills imported South American wool primarily for domestic-market products. Nowadays, however, joint ventures and other special arrangements have made the more successful township-enterprise mills which survived the 1989 to 1991 "shake-out", much more export-oriented. In the 1990s, it is highly likely that much more than 25% of the wool imported into China will be destined for re-export in some processed form.

11.2 Reforming Importing Arrangements

Wool-importing arrangements in China in the first half of the 1990s have reflected a system in transition. The two reforms to attract most attention outside China in connection with the wool trade have been the move to a single exchange rate and the reduction in tariffs. However, the subtle and almost unofficial adjustments both to the institutions concerned with importing wool and to the procedures mills must follow to gain access to foreign supplies, are probably more important than the two widely publicised trade policy reforms.

11.2.1 Exchange Rate and Tariff Policy Reform

China introduced major changes to its exchange rate system and significantly reduced tariffs on wool imports at the beginning of 1994. Traditionally, China operated a dual currency system with local and foreign exchange components. The Central government rationed foreign exchange at an official fixed rate of exchange which usually overvalued the local currency. To combat illegal "black markets" in foreign currency and to overcome the inefficiencies associated with currency rationing, the government established a system of legal secondary markets for foreign exchange in the second half of the 1980s.

The transaction costs of operating in these secondary markets have been identified as impediments both to raw and semi-processed wool imports and to the export of wool-based products (Martin, 1992; Morris *et al.*, 1993).

In 1994, a single currency system was introduced although the extent to which the exchange rate will be allowed to adjust freely remains unclear. At the same time as the exchange rate reform was introduced, China also agreed to reduce the tariffs on semi-processed wool by one-quarter and on raw wool by one-third (although the latter reductions were to be the subject of review in 1995). The impact of these tariff reductions can easily be exaggerated because wool destined for re-export has been imported duty-free since 1992.

Joint ventures and other mills producing primarily for the export market gained little as they already imported wool duty-free and, as exporters, they had access to foreign exchange. On balance, therefore, other factors are likely to be more important in determining future levels of wool imports than the tariff reforms. At the same time, the net impact on wool imports of the changes in the foreign exchange system are difficult to predict because of the uncertainty surrounding the precise nature and extent of the reforms themselves and the economy-wide effects these reforms may have (Martin, 1992). Nevertheless, in terms of the Chinese wool textile industry, the 1994 exchange rate and tariff reforms are likely to have the greatest beneficial impact on mills which have traditionally processed principally for the domestic market, such as the up-country mills in the wool-growing provinces. These mills should now find it easier to obtain foreign exchange, and the lower tariffs will mean that the net cost of wool imported for use in the manufacture of products for the domestic market will be lower than it would otherwise have been. In the negotiations about China's accession to GATT, both further reforms to the foreign exchange system and a reduction in wool tariffs are on the agenda.

11.2.2 Freeing-up the Importing System

The various phases through which the Chinese wool-importing system has passed since the late 1970s have been succinctly described by Wilcox (1994). This section draws on that paper as well as the detailed insights gained during numerous discussions both with Chinese wool industry and trade officials, and with representatives of the IWS/AWRAP in Hong Kong and Beijing. The aim is to demonstrate some of the subtleties of the Chinese wool-trading business.

The second half of the 1980s was a chaotic period for the Chinese wool textile industry. The traditional rigid controls over imports had broken down and many different agencies had entered the wool-importing business with or without official approval. The sudden downturn in the Chinese economy following the harsh macro-economic policies introduced in late 1988 to combat inflation led to a major contraction in wool-trading activity. The Central government seized the opportunity to reintroduce centralised control over wool imports in 1989. However, as the economic situation improved in the early 1990s, pressure grew for major administrative reforms as industry participants recognised the shortcomings of the highly centralised wool-importing arrangements. Chinese officials faced the dilemma either of retaining strict controls and thereby continuing to distort and retard the modernisation and development of the wool textile industry; or of moving to completely free-trading arrangements with the risk of recreating a chaotic trading environment such as that which existed in 1988. In an attempt to steer a middle ground, they have allowed "loopholes" to develop in the apparently rigid arrangements to facilitate growth in imports and permit the industry to develop more-or-less in response to market opportunities. This *de facto* freeing-up of import arrangements exerts different impacts across the wool-processing industry. Some mills are in a better position to exploit the loopholes than others. For example, the effects of these more-or-less unofficial administrative reforms have varied markedly between joint-venture mills and nonjoint-venture operations, between mills producing for the export market and the domestic-market-oriented mills, and between the east coast and up-country mills.

As mills and related textile industry organisations sought ways around the official wool-importing arrangements, they jostled for position and completely new importing channels emerged. This evolutionary (or as the Chinese would say "step-by-step") process will take time to stabilise. In the meantime, foreigners (and many Chinese as well) who wish to be involved in the wool trade with China will find it an extremely difficult and frustrating business. However, as is so often the case in China, the best preparation for understanding current arrangements is to appreciate the recent history of the situation of interest.

As discussed in Section 11.1.2, Chinese wool imports rose dramatically between 1981 and 1988. Traditionally, wool imports were tightly controlled and were part of the State Plan developed by the State Planning Commission. The detailed administration of wool imports involved MOFERT and MOTI. The former was responsible through its national trading corporation—the China Textile Import and Export Corporation (better known as CHINATEX)—for organising the purchase and importation of foreign supplies, and the latter was concerned with the allocation of the available supplies of imported wool to individual mills. However, the sudden growth in wool-processing capacity in the mid- to late-1980s, much of it owing to the development of township-enterprise mills outside the control of MOTI, led to provincial trade corporations being permitted to import wool directly. Initially, these imports were more-or-less part of the State Plan. But just as domestic wool marketing became chaotic in the second half of the 1980s, so did wool importing. Individual mills and local authorities actively competed for access both to a limited amount of domestic wool and to imports. Wool import controls became virtually non-existent, especially in 1988.

As stressed above, the sudden downturn in the market in 1989 allowed the Central government to re-assert its control over wool imports under the auspices of MOFERT. A new organisation, which was to prove to be extremely ephemeral, was established in 1989 to coordinate all wool imports, excluding imports of foreign joint ventures. This so-called China Wool Group consisted of one representative each from seven trading corporations and was chaired by the CHINATEX representative. The seven corporations were CHINATEX; China National Animal By-Products Import–Export Corporation; China International Trust Investment Corporation; Tianjin Foreign Trade Corporation; Beijing Foreign Trade Corporation; Shanghai Foreign Trade Corporation; and Jiangsu Foreign Trade Corporation. In terms of control over wool imports, however, this particular organisation disappeared almost as quickly as it was established and by 1991 was all but defunct as pressures emerged to import outside of this system.

In 1989, the Central government sought not only to regain control over wool imports but also to restrict the level of imports. The quota nominated by the China Wool Group for 1990 of 45,000 tonne (clean-wool equivalent) was well below import levels in 1988 and 1989 (Fig. 11.2). Various reasons have been advanced for the restriction on imports (see, for instance, Wu, 1990). One major reason cited was the level of finished product and raw wool stocks held in China at the time. Raw wool stocks held by the SMCs at the end of 1988 were larger than normal at 68,000 tonne. By the end of 1989, the SMC stocks had reached almost to 120,000 tonne (greasy) or about half the annual domestic production. The SMC stockpile peaked at 143,000 tonne (greasy) in 1990 (Lin, 1993). To finance these raw wool stocks, the Ministry of Commerce provided ¥500 million to the SMCs as an interest-free loan with no fixed repayment schedule (Section 7.3). Reducing imports was considered to be one way of encouraging use of the domestic wool stockpile. The type of domestic wool held in stock, however, differed markedly from imported wool, primarily being low-quality fine wool Grade II and improved fine wool Grade II. Lin (1993) demonstrated that mills did not use the low-quality and relatively high-priced domestic stocks to replace imports, but instead increased the proportion of chemical fibres in their final products.

Recentralisation of control over imports in 1989 meant a return to traditional arrangements. Since wool had always been categorised as a strategic item with respect to importing procedures, the amount to be imported each year (the import quota) was traditionally determined as part of the State Plan. State enterprises (mills) sought permission to import wool from their local Economic Planning Commission and from MOTI. It was the task of MOTI to gather all the orders and make a recommendation to the State Planning Commission which then developed its overall Plan and issued licences to MOFERT to import wool. MOFERT traditionally only authorised its own agencies to import wool, but in the second half of the 1980s, as already described, it initially allowed provincial trading corporations and some of the larger mills to import wool directly. Then, in 1987 and 1988 MOFERT virtually lost control of wool imports, and a large number of government and semi-private organisations became involved in the wool-importing business.

When MOFERT regained control over wool imports in 1989, it reverted to authorising CHINATEX to organise the importation of fine and semi-fine wool and the China National Animal By-Products Import–Export Corporation to handle non-apparel-wool imports. These two MOFERT agencies have existed for over 40 years. (As explained in Section 7.2.4, the China National Animal By-Products Import–Export Corporation and its subsidiaries at the provincial level are not connected with the animal by-product or native goods companies within the SMC structure.) In the past, the Animal By-Products Import–Export Corporation frequently encroached on the territory of CHINATEX by becoming involved in apparel-wool imports. However, in 1989 CHINATEX sought to control all imports of this type of wool.

Three main "autonomous" subsidiary organisations acted for CHINATEX, namely: China Resource Textile Corporation (based in Hong Kong); Nam Kwong (based in Macau); and CHINATEX Sydney (located in Sydney, Australia). These three subsidiaries of CHINATEX had been handling wool imports since 1984. The official rule was that 70% of apparel-wool imports was to be handled by CHINATEX itself and 30% was to be the responsibility of the three subsidiaries. Various industry sources both within and outside CHINATEX, however, claimed that by 1992 much more than 30% of the quota trade was bypassing the parent bureaucracy of CHINATEX.

Indeed, by 1992 importing arrangements had been significantly relaxed once more. In general, any mill wanting to import wool under the quota could apply to do so through the local provincial textile industry corporation (TIC) or foreign trade corporation. These provincial corporations could then negotiate with one of the three CHINATEX agencies mentioned above. Eventually, these agencies purchased the wool either at overseas auctions or from international wool traders. Although import quotas applied to individual provinces, they did not specify wool type or country of origin. Furthermore, in line with the less stringent importing conditions which existed in 1992, the quotas were considered more indicative than regulatory and transfers between mills and textile industry corporations in different provinces were permitted.

More importantly, some mills were able to gain access to wool imports outside the quota system. This is one reason why the MOFERT statistics for wool imports in some years differ from (are less than) the corresponding statistics collected by Chinese Customs. MOFERT only monitors wool imports which are part of the official quota.

The principal loophole which has emerged in the quota system has been the establishment of joint ventures. The incentives for joint-venture arrangements with foreign business units are considerable. Import quotas do not apply to wool imported as part of a joint venture to process the wool for re-export. As already mentioned, imports of this type are also exempt from the tariffs on wool. In addition, most joint ventures enjoy taxation concessions and are often given preferential treatment in other respects by local authorities. It is not surprising, therefore, that wool mills and wool traders in China have embraced the joint-venture concept with great enthusiasm.

The precise administrative mechanism by which officials of the Chinese Customs Service mark the import licences and keep track of joint-venture wool imported with the stated intention that it will be re-exported in processed form is outlined by Wilcox (1994). The logistics of tracking imported wool in China, however, suggest it is a difficult task and the system is open to exploitation. The increasing popularity of joint-venture processing for re-export since 1992 is testimony to the "success" of this loophole in the quota system. Undoubtedly, this almost unofficial easing of import controls has played a major part in lowering the black-market price for imported wool in China (Wilcox, 1994).

Apart from the evolutionary semi-official relaxing of importing arrangements described above, there has also been some major official institutional restructuring, especially in 1993. These changes are outlined in Section 2.1 and, in principle, appear to represent a major downgrading of the bureaucratic power of MOFERT (now MOFTEC) in relation to wool importing and some transfer of authority over the wool trade from MOFERT/MOFTEC to the restructured MOTI. For example, several major trading corporations previously controlled by MOFERT now operate directly under the new MOTI. Two of these, namely the China Textile Resources Company and Beijing Xie Li Textile Limited Corporation, are now major wool-trading companies. For example, Beijing Xie Li Textiles received an official import quota/licence for 5,000 tonne in 1991, which was around 4% of the total wool import quota in that year. This company has an office in Sydney where it is known as the "Three Rings" Company. It buys in the Australian greasy-wool auctions through one of the major Japanese wool-buying firms. Beijing Xie Li Textiles enters into joint ventures with textile mills in many provinces but especially in Jiangsu, Tianjin-shi and XUAR.

The relaxation of the import arrangements in the 1990s and the various loopholes which have arisen in respect to licensing have led to a multitude of diverse organisations with direct authority to either import or deal in overseas wool (Wilcox, 1994). Apart from the traditional Chinese wool-trading corporations (such as CHINATEX and its three subsidiaries), some provincial foreign trade corporations, intermediate wool-supply companies, and even some mills now have authority to import wool directly. Apart from these organisations with the authority to import wool, various other agencies also deal in imported wool. For instance, there are intermediate wool-trading corporations which virtually act as imported-wool wholesaling agents by arranging large orders of imported wool. These traders satisfy the needs of mills by being able to supply wool in small quantities and on a regular basis. Their access to finance, storage and handling infrastructure, and contacts with State authorities, enable them to operate in such a manner.

At the extreme end of the new entrepreneurial activity concerned with wool imports are a plethora of small trading companies which operate in the secondary wool markets. These companies have no direct authority to import but purchase wool from approved importers. The continued push for open-market policies, along with developments in and access to new communication technology, has further developed these markets. However, with the rapidly expanding opportunities for individuals and small companies in China, these markets may quickly become chaotic. Wilcox (1994) outlines attempts to provide some market infrastructure by establishing places for these dealers to operate in Nanjing, Zhangjiagang, Changshu and Jiaxing, and by plans to create such markets both in Kunshan and in the new Pudong special economic zone in Shanghai.

The emergence of many different Chinese agencies with the capacity to import wool creates new opportunities but also new uncertainties for overseas wool exporters. Previously, exporters could focus their attention and activities on CHINATEX or its agents. But now they must deal with many new, lesser-known participants. Transitional problems with the more open importing system emerged as early as 1992. Power (1993), citing the Australian Council of Wool Exporters and various wool brokers, claimed that wool contracts worth AUD$25 million, accounting for 6% of Australia's wool trade with China, had been broken. The breaking of these contracts was linked to individual mills and textile companies dishonouring or trying to re-negotiate contracts following a fall in international prices. The response of Australian exporters was to address the problems at a diplomatic level by having the issue raised on a government-to-government basis. Such an "old-fashioned" approach would seem to be inappropriate and ineffective in the new Sino-Australian wool-trading environment of the 1990s. A more lasting resolution of the problem, given the new less-centralised trading environment, calls for more direct contact at the grass-roots level. The success of the recent reforms, both for Chinese mills and for overseas exporters, depends critically on forging closer links and greater understanding between the Chinese mills and overseas suppliers.

11.3 Up-Country Mills and Importing Wool

Various levels of government seek to promote further development of up-country textile mills. Authorities in wool-growing provinces, in particular, perceive that an integrated wool industry can play a key role in the economic development of these remote areas which have few other industrial opportunities open to them. The Central government is especially sensitive to the need to improve economic welfare in these strategic and sometimes politically volatile parts of northern and north western China. Traditionally, the up-country mills in these provinces have been expected to give priority to processing locally-grown wool. Indeed, the importance of the wool textile industry in XUAR, IMAR and Gansu, derives not from its share of national processing capacity but from the fact that these three provinces grow over two-thirds of the fine wool (≤25µm average fibre diameter) produced in China.

Imported wool has been perceived by some as a threat to the domestically-grown wool, resulting in pressure to exclude wool imports from GATT negotiations. But if the up-country mills are to adjust to the new competitive pressures resulting from a freer market for wool and wool-based products in China, they need imported wool and foreign technology to counter the deficiencies (quantity and quality) of locally-grown wool.

Textile mills located on the east coast have significant advantages over inland mills in relation to processing imported wool for re-export. However, these advantages are likely to be less significant when the imported wool is processed for the domestic Chinese market. Of the three sectors in the Chinese wool textile industry identified in Section 2.3.3, the up-country sector has been the most isolated in relation to foreign trade. These mills have traditionally produced for the domestic market, with little opportunity to import wool or to export their products. Many of these mills are located in the 12 provinces which together make up China's pastoral region (Longworth and Williamson, 1993). The mills in XUAR and IMAR (the two major wool-growing provinces) and in Gansu (another significant wool-producing province) are of particular interest. Together, the mills in these three provinces represent a little less than 10% of all wool spindles in China (Section 2.3.4). More details on the up-country mills, including the problems that confront them and their potential and need to import further wool, is provided in Brown and Longworth (1994).

To gauge the extent to which the up-country mills have been "starved" of imported wool in the past, consider first the case of the XUAR. During the first half of the 1980s, the XUAR had a modest import quota of around 500 tonne with a peak of 900 tonne. The quota primarily was used to purchase New Zealand coarse wool (35–38µm) for the production of hand-knitting yarn. However, imports of up to 300 tonne per annum of Australian fine wool also occurred. The "self-produce, self-process and self-sell" policy, introduced in 1985 and discussed in Section 2.3.4, further eroded these modest imports since there were no import quotas allocated to XUAR between 1985 and 1991. Nonetheless, around 200 to 300 tonne of imported wool per annum was used by XUAR mills. The imported wool was mainly purchased from other mills or provincial TICs through the domestic market, but on some occasions the XUAR TIC exchanged XUAR wool, generally of higher quality than other Chinese wool, for imported wool held by other provincial TICs. With an easing of the self-sufficiency policy and trade restrictions in 1992, XUAR was again allocated an import quota of 600 tonne for use with top-quality domestic wool in the production of products for export. The XUAR's export effort has been hampered in the past by the self-sufficiency policy, with exports accounting for less than 1% of total textile output.

The "self-produce, self-process, self-sell" policy also resulted in minimal import quotas for the IMAR between 1985 and 1991. Most of the imported wool which became available was used in the production of export products, with the remainder going to domestic worsted fabrics and hand-knitting woollen yarns (using 48/50 count wool). Officials in the IMAR textile industry claimed that they had little say in their quota allocations and that they were actively discriminated against under the self-sufficiency policy. The limited quota wool assigned to the IMAR had to be imported through CHINATEX. In addition, some wool imported by MOTI was made available to the IMAR by the Tianjin branch of the Beijing Xie Li Textile Limited Corporation. For access to additional imports, IMAR mills had to rely on imported wool offered for resale by mills in eastern and southern China. As elsewhere in China, joint ventures such as the Sino-Japanese Green Pine factory and the Green Bean Textile Corporation in Huhehot, the capital of IMAR, have been developed to circumvent the policies which restrict access to wool imports. As in the XUAR, however, experience with the foreign trade sector is limited as compared with some of the eastern provinces. Thus while joint ventures have facilitated a small amount of exports of wool products from the IMAR in the past, further development of an export trade has been constrained by the inability of the mills in IMAR to gain access to foreign supplies of wool.

With respect to wool imports, Gansu differs slightly from the other major pastoral-region provinces. Being a less important wool-producing pastoral province than XUAR and IMAR, and with a long history of textile processing, Gansu has enjoyed relatively more favourable access to imports. Nonetheless, imports of around 1,000 tonne clean represent only one-fifth of the total demand for wool by Gansu mills. The source of these imports, however, varies from year to year. In 1989, 600 tonne of clean wool was purchased on the domestic imported-wool market, mostly from intermediate traders. Most of the imported wool is used in the production of export products, and 70 to 80% are pure-wool products. For Gansu No. 1 mill, one of the largest mills in the province as well as in China as a whole, imports usually account for 25% of the wool it purchases. Exports account for around 20% of the mill's output and are sold principally to Japan, Hong Kong and some South East Asian countries in the form of worsted and woollen cloth, blankets, and knitting wool to be made up into garments such as jumpers. In the past, mills have been able to retain up to 80% of the value of these exports to purchase overseas wool or foreign equipment and parts. In 1992, mill managers in Gansu regarded the lack of foreign exchange, rather than access to foreign supplies, as the major constraint to further wool imports. The agency responsible for wool imports, the Gansu TIC, typically imports through CHINATEX. Other small wool-importing companies exist in Gansu but these companies have also operated through CHINATEX in the past.

As explained in the previous section, the general rules under which import quotas are administered have been relaxed significantly since 1992. Nevertheless, mills seeking access to imports for production aimed at the domestic market still face major obstacles to obtaining both the type and quantity of imported wool they require. Most of their supplies must still be imported under the quota system through official channels. Some imported wool may be purchased from other mills, notably in the coastal and southern regions, which were allowed to import wool in excess of their requirements (perhaps because they deliberately overstated the quantity of raw material required to satisfy their export orders). Alternatively, some export-oriented joint-venture arrangements permit the mills involved to use up to 30% of their wool imports for the production of domestic-market products without jeopardising the generous concessions they have been granted. These joint-venture mills sometimes subcontract out to other mills the production of domestic-market products from their imported wool. Another common ploy, especially in the case of township-enterprise joint ventures, is for the arrangement with the foreign joint-venture partner to apply to only one production line in the mill. This makes it easier for the mill to divert surplus imported raw materials to other production lines producing for the domestic market.

However, for the up-country and other primarily domestic-market-oriented mills, the major source of imported wool remains the more-or-less centralised official quota system. The major problems this creates for these mills stem from the length of the ordering chain.

11.4 Importing Channels and Information Flows

Earlier it was stated that under the official quota arrangements, mills allocated a share in the import quota place their orders with their provincial textile industry corporation or foreign trade corporation, and that these corporations must deal with CHINATEX or one of its three subsidiaries, etc. When the centralised system was strictly enforced, mills had to place their orders nine months or more in advance. Even in the post-1992 environment, mills still complain of the excessive lead-times associated with the official wool-importing channels (Brown and Longworth, 1994).

Not only are the official importing channels slow to deliver, they also lose information along the way. Mills commonly specify their requirements in terms of fineness, length and vegetable matter fault. One of the authors interviewed a significant number of mill managers in IMAR, Gansu and XUAR in 1992. The overall impression gained was that these managers were not well informed about the latest additional measurements and specifications being used in the international wool trade. Officials of the provincial TICs (the quasi-independent peak industry organisations representing the mills in each province) were, in general, better informed. The TICs collect the orders from the mills and at that stage can "advise" the mills about the specifications in the orders. Once the orders are forwarded to CHINATEX, they are lumped together and the specifications are sometimes changed before the final purchase orders are placed with an importing agency.

While the arbitrary adjustment of the specifications in their orders somewhere in the importing chain was clearly a major problem for the inland mills, they also expressed considerable dissatisfaction with other aspects of foreign wool. For example, according to the Gansu TIC, at least one-third of the Australian greasy wool imported into Gansu was supplied at the lower bounds of the specifications in the purchase order. This problem was compounded by the broad quality bands implicit in the orders in the first place. The Gansu mills also complained about the substantial intra-lot variability in the imported wool. As a result of the problems encountered with Australian greasy wool, particularly in relation to the variability of fibre length and vegetable matter content (burrs) in 1988 and 1991, the Gansu No. 1 Mill reverted to purchasing tops from Australia. Since other mills are likely to have encountered similar problems, this probably explains the increase in the proportion of Australian wool imported by China in the form of tops which was referred to in Section 11.1.2.

The question of the amount and type of vegetable matter in Australian wool is a good example of the need for better communication between overseas suppliers and Chinese mills. In general, while Chinese-grown wool often contains a great deal of foreign matter (dust, sand, faeces, leaves, etc.), it is remarkably free of burrs and grass seeds. On the other hand, Australian greasy wool, especially wool from the drier pastoral areas of the country, often has a relatively heavy load of this kind of vegetable matter. While these impurities can create problems for Australian mills (see, for instance, Francis, 1992, p.51), the scouring and carbonising techniques employed in China are even less well suited to processing wool containing noogoora, bathurst and galvanised burrs, spear grass, barrel medic, bogan flea and burr medic. The broad nature of the burr in the wool is one of the specification details available to buyers of

Australian wool, but this information is often "lost" somewhere along the way to the inland mill in China. This problem has attracted the attention of the Director of the Australian Council of Wool Exporters, who has argued that there needs to be more objective testing information provided about the precise type of vegetable matter in each lot (Francis, 1992, p.53).

The lengthy and poorly functioning importing channels in China work against the spread of market information about imported wool. Although CHINATEX and its agents disseminate some information through the local TICs, the information is often limited and invariably outdated. The past policies which isolated the up-country mills from the foreign trade sector mean that many of these mills, including the more innovative ones keen to import wool, remain ignorant about important aspects of the international wool markets, including potential alternative purchase channels, price and auction information, wool grading and testing, and general market conditions in the exporting countries. For instance, a number of mill and other textile industry officials interviewed in the second half of 1992 believed that Australia still operated a reserve-price/buffer-stock scheme for wool. The abolition of this scheme early in 1991 was an epoch-making event in the history of the world wool trade, with profound implications for Chinese importers. That many key people in the Chinese wool textile industry were unaware of the demise of this scheme over a year after the event, graphically illustrates the lack of knowledge which exists in China in relation to the international wool market.

The recent emergence of many traders and dealers in imported wool in China may not necessarily alleviate these information problems and, under given circumstances, may worsen the situation. Wool may be exchanged a number of times on the secondary market before actually reaching mills, and in the heat of the market, detailed specifications are often ignored or lost. Under such conditions, Wilcox (1994) argued that imported wool tends to be branded in generic terms and test certificates and shipping documents tend to be incomplete, offering great scope for fraudulent transactions. Thus the appropriate checks and balances need to be placed on these markets if they are to serve the important function of linking the demand preferences of Chinese mills and consumers with the production plans and supply characteristics of overseas suppliers.

Some of the recent changes in the wool import arrangements may improve the information flow. Officials of the Gansu TIC, for example, reported with enthusiasm that in 1992 they had been able to obtain market information from the Beijing Xie Li Textile Corporation as well as from CHINATEX. But mills need to be informed in much greater detail and on a much more timely basis if they are to make the most efficient use of wool imports. Specifying requirements based on outdated and limited price information for purchases arriving six to nine months later is not an appropriate basis for good decision-making. Mills interested in importing wool need more direct access to auction catalogues and other information about overseas markets.

Although more direct marketing channels may be desirable, they are not necessarily without costs. There are likely to be significant economies of size for the Chinese in centralised wool importing, in the provision of market information and in conflict resolution. More importantly, a centralised importing system may avoid the excesses of destructive, parochial, inter-regional competition for imports. Nevertheless, the centralised arrangements of the 1980s and early 1990s fell well short of the needs of mills to the extent that they severely curtailed both the level and efficient use of imports.

In the move to more open trading arrangements, however, new institutions are needed to support these arrangements. To reap full benefit from the reforms, Chinese decision-makers need to facilitate and improve market channels, information flows and other service areas to enable individual mills to make appropriate purchase decisions and to operate in the more open, but potentially more chaotic, trading environment. This is especially critical for pastoral-region mills which, because of their location and history, have had less direct exposure to the foreign trade sector. Changes in the role and functions of the Ministry of Textile Industry (Section 2.1) may represent a modest move in this direction.

11.5 Concluding Remarks

China is now a dominant player in the world wool market both as an importer of raw and semi-processed wool, and as a consumer and exporter of wool-based textile products. Progress in multilateral trade negotiations and further growth in Chinese household incomes will reinforce these developments. The recent remarkable dynamism in the Chinese wool textile industry has attracted the attention of agribusiness entrepreneurs throughout the world, with the Chinese wool trade being seen to offer significant new commercial opportunities. However, foreign firms "seeking a slice of the action" in this dynamic though highly volatile market need to invest heavily in understanding all aspects of the Chinese wool scene if they are to be successful.

Liberalisation of wool trading in the early 1990s has been aimed at improving the efficiency of importing channels and information flows. The changes introduced have also been designed to encourage the re-export of wool-based products. However, many mills, and in particular the up-country mills, produce primarily for the domestic market. As the Chinese economy grows, so will the demand for wool-based goods. The Chinese domestic market is widely acknowledged both by Chinese and international experts as having an enormous growth potential (e.g. Zhang, 1990b; Connolly and Roper, 1991; Garnaut *et al.,* 1993; IWS, 1994). The Chinese wool-growing industry was investigated by Longworth and Williamson (1993) who concluded that there is relatively little scope for China to produce significantly more high-quality apparel wool. If the potential demand for better-quality wool-based products in China is to be met, the mills producing for the domestic market will need access to greatly increased amounts of apparel-type wool from overseas.

In order to make efficient use of wool imports, or even to realise the profitability needed to purchase the imports, the problems facing the Chinese processing sector highlighted in Chapter 10 need to be overcome. Many of the factors adversely affecting Chinese wool processing and subsequently the demand for imported wool can only be resolved within China. However, there are a number of measures that overseas wool exporters can adopt to assist the importing process.

The domestic-oriented mills such as those in the pastoral region need to be approached differently to the export-oriented mills of the east coast with which foreign suppliers are more familiar. Pastoral-region mills are being weaned off policies which have insulated them from overseas markets, and they face considerable transition problems and are in desperate need of better marketing information. Their insulation from foreign markets means that they are largely ignorant of overseas markets, especially recent developments, and their official and unofficial networks for obtaining that information are poor. Another area where overseas exporters can usefully facilitate the import process is in the provision of technical assistance, not only in the more obvious areas of spinning and scouring technology (Chapters 9 and 10) but also in other crucial fields such as fibre inspection and testing (Chapter 6).

Some innovative steps to facilitate wool trade have already been taken both by major Chinese institutions and by foreign organisations interested in promoting exports of wool to China. For example, the restructured MOTI is expected to service the needs of the Chinese wool industry by providing the latest technical and market information. It is being assisted in these endeavours by outside help from the IWS and AWRAP. These centralised initiatives need to be encouraged. However, the heterogeneous nature of Chinese wool textile mills and their dispersion across the country create a large number of potential market niches for foreign suppliers and joint-venture partners. These opportunities can be identified with the aid of industry-wide support agencies such as MOTI or IWS/AWRAP. Ultimately, however, the extent to which these opportunities develop into mutually worthwhile business arrangements will largely depend upon the willingness of each side to understand and to adjust to the trading needs of the other.

Chapter 12

Lessons for the Future

The opening up of the Chinese agribusiness sector has created an enormous range of new opportunities for foreign interests. However, as the experiences of foreign entrepreneurs in other sectors of the Chinese economy have demonstrated, converting agribusiness opportunities into profitable business ventures will not be easy. Commercial success requires an appreciation of the recent history of the market, of the rules and regulations governing the market, and of the organisations and institutions participating in the market. This book provides this information in relation to the wool market and also demonstrates the complexity and sophistication which exists in the Chinese agribusiness sector in general, and with which any potential new entrant needs to be familiar.

Thus, this volume is not just about "soft gold" but, as is suggested in the opening pages, it provides a "window" through which to view not only the Chinese wool-marketing system but also a much wider spectrum of "things Chinese". The need to appreciate the social, cultural, economic and political milieu in which the marketing of wool is embedded is a recurring theme in the text.

To understand the complexity of this agribusiness environment and to appreciate the nature and impact of recent reforms designed to improve the functioning of the wool market, requires a detailed treatment of the subject such as that found in the preceding chapters. At the same time, this book provides many general insights into the economic reform process in China. To date, the transformation of the Chinese economic system from a centrally-planned command economy to a socialist market economy has progressed more slowly but with better results than the "quick-fix" approach adopted in many parts of the former USSR and Eastern Europe. The wool-marketing case illustrates well the advantages and disadvantages of the Chinese step-by-step approach to economic reform. Both in this context and in other respects, there are many rather general lessons for the future which emerge from this study. The remainder of this chapter briefly discusses four of these major conclusions.

12.1 A Mega-Agribusiness Sector is Emerging in China

As the world's largest producer and consumer of agricultural products, China also potentially has the largest agribusiness sector. The economy-wide rural reforms of the post-1978 era initially stimulated remarkable production responses, with total agricultural output growing at 7.1% per year between 1978 and 1984. Even when the growth in output slowed in the second half of the 1980s, the expansion continued at an annual average rate of 4.1% between 1984 and 1989 (Shaffer and Wen, 1994).

The generic reforms which provided the framework for these remarkable changes have been well documented. (See, for example, Watson, 1983; Parish, 1985;

Longworth, 1989; Lin, 1992; Shaffer and Wen, 1994.) On the other hand, there have been few detailed studies of the industry-specific reforms which provide the flesh for the broad reform framework, with Chen and Buckwell (1991) being a notable exception. However, studies such as that reported by Chen and Buckwell have concentrated on industry policy issues rather than the detail of reform-induced changes both in the marketing system and in the overall agribusiness environment of the industry in question. In contrast, this book focuses primarily on the detail of these reform-induced changes.

Gradual deregulation of the agribusiness sector has been one of the major outcomes of rural reform in China. From 1992, virtually all agricultural input and product markets have been deregulated to varying degrees, thus opening the way for rapid commercialisation of the Chinese agribusiness sector. This book presents a case study of how one specific part of this sector has evolved. While the specifics are, of course, limited to wool marketing, there is considerable generality in the analysis. For example, the transformation of the wool-marketing system demonstrates three major features common to the reform process in the Chinese agribusiness sector. The first major feature has been the shift from State procurement of production quotas at fixed prices to a competitive market in which, at least in principle, prices are determined by the forces of supply and demand; second, the relatively smooth transition to a market-based exchange mechanism has required that the necessary facilitating trading standards and associated institutions are in place; and third, the principal participants in the new marketing system are still jockeying for position in the new agribusiness environment.

12.2 Wool is Becoming More Important for China

The raising of sheep and other grazing ruminants for meat and animal-fibre production has, until recently, received a low priority at the Central government level. The paramount policy objectives of food security and food self-sufficiency have meant that, traditionally, food-grain issues have dominated political debates about agricultural policy in China. Food-grain production and agriculture generally have been given a higher priority than animal-husbandry activities. Within the animal-husbandry sector, intensive livestock production involving pigs and poultry, and recently dairying, has received the most attention.

Even in the major pastoral provinces such as XUAR, Gansu and IMAR, where the relative emphasis on grazing animals and their products is much greater, food production and agriculture generally take precedence over the production of animal fibres such as wool, cashmere, camel and goat hair.

Despite the relatively low priority traditionally attached to sheep and wool production and marketing in China, there have been significant public investments in the modernisation of the Chinese sheep and wool industry. In particular, there has been a longstanding Central government policy to upgrade the Chinese sheep flock towards fine wool production. The aim has been to increase the quality and quantity of apparel wool available for the Chinese wool textile industry. Nevertheless, in the past, the domestic wool-growing industry has not been an activity which attracted much interest in the "halls of power" in Beijing.

A number of recent (and not so recent) events have combined to raise the status of wool in Chinese policy-making circles. For strategic, political, commercial and trade-policy reasons, wool has now moved to centre stage in Beijing.

The minority nationalities which produce wool as their major source of cash income in the remote pastoral areas have become increasingly sensitive to being "left behind" as China modernises. These peoples have historical and ethnic links with the populations of the surrounding countries and many of those nations are facing a period of great political and economic instability. The Government of the PRC recognises that one response to this new politico-strategic environment is to support a vigorous wool-growing and related wool-processing industry in China's north and north west. The production and sale of wool is one of the few viable economic activities available to large numbers of the traditional inhabitants of China's pastoral region.

In a commercial sense, the remarkable expansion of the wool textile industry is one of the success stories of post-1978 China. Nevertheless, the economic recession in 1989 and 1990 hit the industry extremely hard. Conditions in this important industry were so bad at the beginning of the 1990s that, as discussed in Section 10.1, the Central government singled it out for special attention in the Eighth Five-Year Plan (1991–1995). However, after some major restructuring, the Chinese wool textile manufacturing industry entered a new period of expansion from 1992 onwards.

Given the severe resource and other constraints limiting the expansion of domestic wool production, the growth in the demand for wool inputs generated by the expanding wool textile industry could only be satisfied by imports. China is now a dominant player both in the world markets for raw and semi-processed wool and in international wool textile, garment and yarn markets. The wool textile industry is a significant part of the Chinese foreign trade sector. Indeed, wool trade issues have become central in relation to China's attempts both to rejoin GATT and to be accepted as a founding member of the proposed new World Trade Organisation. The politico-strategic need to protect domestic wool growers makes it difficult for China to relinquish the right to limit wool imports by quota in order to re-enter GATT. On the other hand, being a full member of GATT would greatly assist China to overcome foreign restrictions on its wool textile exports.

Not only is wool a commodity of much greater significance to China in the 1990s than at any time in the past, but it is also a commodity which is likely to continue to require the attention of Chinese policy-makers for the foreseeable future.

12.3 China is Becoming More Important for Wool

China was an extremely minor participant in international wool and wool textile markets prior to the 1980s. Yet, in little more than a decade it has emerged as the single largest trader in these markets. As pointed out at the end of Section 11.1.2, while it is difficult to obtain reliable statistics, it has been estimated that on average only about 25% of the wool imported by China is re-exported in a processed form. However, in the past, the proportion re-exported in any given year varied greatly as the amount of wool imported changed abruptly from year to year. Once China rejoins GATT, Chinese exports of wool textiles, garments and yarns are likely to become even more dominant in world markets.

Since wool products are essentially luxury consumer-durables for most people in China, the consumption of wool within China increased rapidly as incomes rose during the 1980s. Domestic demand contracted sharply in 1989, initially as a consequence of the strong anti-inflationary measures introduced in late 1988 and then also in response to the political uncertainties created by events in the first week of June 1989. As the Chinese economy recovered in the early 1990s, so did the domestic demand for wool-based products. Both Chinese and foreign researchers predict that the domestic demand for wool will continue to expand and that Chinese consumers will represent the largest single national market for wool-based goods early in the next century.

With the anticipated growth in both Chinese wool textile exports and the domestic market for wool products, the upward trend in Chinese imports of raw and semi-processed wool is expected to continue. Of course, while it is difficult to estimate the extent of the substitution, Chinese exports will tend to displace wool-based products traditionally produced elsewhere. Consequently, the expansion in Chinese exports will not necessarily increase the total world demand for wool textile raw materials to a commensurate degree. On the other hand, the expansion in the domestic Chinese market will translate directly into an increase in the derived demand for raw and semi-processed wool on international markets. The future growth in the Chinese domestic market is, therefore, of paramount importance to the future of wool as a textile raw material.

There are some major impediments to the future development of the Chinese domestic market for wool products. Perhaps one of the most important concerns is the extent to which mills producing primarily for the domestic market still have only limited access to imported raw materials. In the past, Central government policies and the administration and distribution of wool import quotas were all biased strongly in favour of the large State-owned mills located in the eastern provinces. These mills were encouraged to process imported wool for re-export. With the spread of the joint-venture concept and the general loosening of the administrative controls over wool imports since 1992, many more mills (both State-owned and township enterprises) have gained access to imported wool. However, despite the generally freer access to imported wool, major policy measures are in place which provide strong commercial incentives for the mills to process imported wool for re-export. As the domestic market for wool products grows, so will the commercial incentive to process imported wool for the local trade. In this respect, it is important that the domestic-market-oriented mills, especially those located up-country, upgrade their equipment and human capital so that they can make the most of good-quality imported wool. These mills also need much better access to information about the international markets for raw and semi-processed wool to enable them to participate more effectively in the wool-importing business. If the potential of the domestic consumer demand for wool products in China is to become a reality, foreign wool interests may need to assist the domestic-market-oriented mills to overcome these problems.

12.4 Things Are Not Always What They Seem

The generally optimistic comments in the first three sections of this chapter about the future of agribusiness in general, and wool in particular, need to be tempered for at least two rather different reasons. First, it is widely recognised that continued political and economic stability in China is not guaranteed. While Deng Xiaoping has long since lost the last vestige of real power, his death will represent a major symbolic turning point in modern Chinese history. The potential danger for national stability of the political realignments which are likely to follow Deng's death could be exacerbated by economic problems arising from the need to stem inflation in the second half of the 1990s. The experiences of the 1989 to 1991 period demonstrate how sensitive the fortunes of the Chinese wool market and textile industry are to political uncertainty and economic recession.

The second reason why an unconditional optimistic outlook for the Chinese agribusiness sector and especially the wool market is not warranted is that things are not always what they seem in China. The phenomenon which Longworth and Williamson (1993, pp.20–21 and 321–322) have termed "policy mirage" is a serious problem not only for the unwary foreigner but also for the Chinese themselves. Frequently, after lengthy debate a policy will be developed at the national level and the Central government laws, regulations, etc. necessary to implement the new initiative will be carefully drafted and eventually proclaimed. Interested foreigners will be shown these documents and Central government officials responsible for the activities with which the policy is concerned will insist that the policy is being implemented throughout China. Indeed, foreigners may find it difficult to find anyone in Beijing who does not regard the policy as being in place. Yet, at the grass roots, economic agents go about their business in much the same way as before. The policy is a mirage.

The deregulation of the agribusiness sector exhibits elements of the policy mirage phenomenon. For example, in Section 5.2 it was pointed out that the opening of the wool market has not created a free market—the playing field is far from level. Virtually every segment of China's agribusiness sector has in the past been dominated by large State monopolies. These organisations served as State input-supply and product-procurement agencies under the unified production and distribution system. The SMCs, described in detail in Chapter 7, represent one of the most important such monopolies but there were similar organisations which enjoyed exclusive trading rights in relation to seeds, meat, fuel, food grains, etc. All of the State monopolies were organised like the SMCs as vertical bureaucratic hierarchies with primary allegiance up the chain of command from the township level to the Central government level. At the same time, the linkages between these State monopolies and local governments at the grass roots, as exemplified by the SMCs and discussed in Section 7.1, were often strong. These State input-supply and product-procurement agencies still dominate the agribusiness sector. They have extremely large marketing networks capable of serving the needs of farmers, and they retain their political links with government at all levels. As in the case of the SMCs in relation to the deregulation of the wool market, these agribusiness giants are awakening. Their existing network of service, distribution and collection points, together with their enormous pool of experienced staff, gives these organisations immense commercial advantages relative to any new entrants. At the same time, they have the opportunity to exploit their political linkages to protect their commercial interests.

In many situations in the agribusiness sector, the major competition for the traditional monopoly has come from trading companies established by other Ministries. For example, the Animal Husbandry Industrial and Commercial Company (AHICC), established in the early 1980s by the Ministry of Agriculture, has emerged as a major buyer of wool, especially in XUAR where it now controls a major share of the market (Section 7.4.1). New trading companies such as the AHICC also have access to pre-existing State networks (e.g. the AHB network in the case of AHICC) and to a political power base (e.g. the Ministry of Agriculture). Consequently, they have the capacity to compete with the traditional trading monopolies more-or-less on equal terms. Any foreign firm endeavouring to enter the Chinese agribusiness sector would be well advised to work closely either with a traditional monopoly such as the SMC or with one of the new trading companies operating under the auspices of a relevant Ministry. Despite the rhetoric in Beijing (and the western media), most segments of the Chinese agribusiness sector, while no longer the exclusive preserve of a single organisation, are still dominated by large State-controlled agencies. While the markets for products such as wool are "open", they are far from "free" in the western sense of the word.

Market deregulation is a general policy which clearly has a large element of policy mirage associated with it. But even specific measures designed to reform one aspect of the marketing of a particular product may exhibit the same phenomenon. For instance, the new Raw-Wool Purchasing Standard discussed in Section 4.3 and supposedly introduced throughout China as from December 1993 is not being applied in many areas. For instance, prefectural-level wool officials in eastern IMAR who were interviewed in mid-1994 initially denied that a new standard was supposed to be in place but eventually admitted it was not being implemented in their area because it was "impractical". In some localities, the new standard may be genuinely impractical for technical reasons. However, Central government policy initiatives may be modified or ignored at the local level for other reasons. With the rapid turnover of managerial positions, local officials in China tend to have short-term planning horizons. Thus, Central government edicts with benefits envisaged in the longer-term may be given a low priority by local officials or even completely ignored especially if the measures concerned involve immediate costs. The issue of local autonomy and short-term planning horizons also has implications for overseas assistance. Local officials will seek projects which involve the installation of new equipment or some other tangible immediate benefit. Other less tangible forms of foreign assistance needed for the longer-term development of the industry, such as improved grading systems and information networks, are viewed with much less enthusiasm.

Chinese politicians and bureaucrats have long acknowledged the differences between policy and reality in China. Deng's widely publicised tour of the southern provinces in 1992 and his subsequent support for even greater economic freedom to foster growth is an example of a Chinese leader "going to the countryside to see what is really happening". But the problem of policy mirage is not dependent on geographic distance for its manifestation. The April 1993 restructuring of Central government ministries has created a policy mirage in the national capital. In the context of the wool trade, the apparent changes to MOFERT and MOTI were profound. Yet, for those dealing with the supposedly reconstituted bureaucracies, it has been "business as usual".

While governments throughout the world are plagued by what is referred to above as "policy mirage", China is such a large and diverse country in geographical, cultural and economic terms that Chinese policy-makers sometimes appear to deliberately resolve a problem by creating a policy mirage. This proclivity has often caught out unwary foreign researchers, trade negotiators and agribusiness operators alike, but those foreigners aware of this predisposition of Chinese policy-makers may be able to take advantage of it. For instance, the debate concerning import quotas for wool calls for a political rather than an economic solution. That is, even though this book has illustrated that imported wool is necessary to maintain the viability of up-country mills which process the bulk of the locally-grown wool, the Government of the PRC does not want to be seen, albeit incorrectly, to be sacrificing the interests of the minority nationalities who produce wool in China as part of the price for rejoining GATT. The most attractive solution for Chinese policy-makers is to create a policy mirage. That is, wool import quotas will nominally be set to appease domestic political concerns. However, the quotas will be administered, or perhaps more descriptively not administered, in a way that allows a level of imports which meets the economic interests of China as a whole, its minorities, and its trading partners within GATT. For better or worse, however, foreigners must operate with these policy mirages. To take full advantage of them requires a thorough understanding of the underlying reality.

So often in China reality only emerges after repeated questioning. There is no substitute for first-hand fieldwork-based research if the true situation is to be appreciated. Desk studies based on secondary sources are rarely likely to yield reliable information. Furthermore, glib generalisations can prove most misleading since the detailed reality is often quite different to what a broad overview may suggest is the case. This book presents a detailed case study of one important facet of the Chinese agribusiness sector which is based on extensive fieldwork in order to separate reality from mirage.

12.5 Final Statement

Napoleon and many other Western leaders over the last few centuries have pointed to China as a sleeping giant and emphasised the potential impact on the rest of the world should the giant awake. Since 1978, China has indeed awoken and it is now seen as rapidly becoming the economic engineroom of Asia. The next few decades could see a profound shift in the world's economic and political centre of gravity from Europe and North America towards China and the other rapidly developing countries in north, east and south east Asia. The one-quarter of humanity which lives in China has only just begun to make its presence felt in many international arenas, but within a generation it could dominate many world markets and political forums.

The transformation of China, therefore, is of profound significance to the rest of the world not only because of what is being achieved in China and its implications, but also because of the lessons which can be learnt and applied in other countries. A great deal has been written in general terms about the post-1978 reforms which have initiated China's transformation. However, there have been few studies which carefully plot the course of change for a specific industry or market.

Agribusiness is not a sector of the Chinese economy which has attracted much attention from foreign analysts. Yet, as stressed earlier in this chapter, the Chinese agribusiness sector is potentially the largest in the world. Similarly, wool is not a commodity commonly associated with China but, in relation to the international wool scene, China has already achieved a dominant role. In a sense, the emergence of China as the key player in world wool markets could be a harbinger of things to come in many other commodity markets.

At this time, therefore, there is an urgent need for detailed studies of how China is transforming specific parts of its agribusiness sector such as the wool market. This volume presents one such case study.

APPENDICES

APPENDICES

The (Old) National Wool Grading Standard

(Issued by the State Standards Bureau in 1976 and implemented on trial
throughout China since then)

The National Wool Grading Standard (i.e. the raw wool purchasing standard) was initially drawn up in 1957. Since then, wool production and wool quality have changed greatly owing to the development of sheep production and the wool textile industry in China. The original wool grading standard developed in the 1950s could not accommodate these developments. Consequently, in 1975 the State Standards Bureau in collaboration with a number of other concerned departments and bureaus began to develop a revised set of grading standards.

(I) The importance of a wool grading standard and principles on which it should be based

Raw wool is a material which has traditionally been handled exclusively by the State in China. The formulation of a wool grading standard should encourage the majority of farmers and herdsmen to increase wool production; it should promote an improvement in sheep breeding; and it should enhance the continuous improvement of wool quality. It should also promote the development of the wool textile industry as well as facilitate raw wool purchasing in the domestic market. Only when the combined initiative of the majority of farmers and herdsmen has been mobilised will more and better-quality wool be produced, thus enabling China to become more self-sufficient in regard to raw wool.

The principles for the formulation and revision of the Wool Grading Standard include the need to allow for advances in the measurement of technical indicators of wool quality; the need to recognise the economic characteristics of wool; and the need for the adoption of a nation-wide viewpoint. Only by negotiation will all interested parties accept the revised standard. It is important to avoid the situation in which every organisation acts according to its own rules without a unified nationwide standard. It is essential to consider fully the opinions and requirements of each department concerned and the feasibility of implementing the standard in the wool-using departments, as well as the opinions of the farmers and herdsmen and of the purchasing departments. Therefore, a thorough investigation should be conducted, taking the international advanced technical standard as a reference point. The domestic production level and technical sophistication should be considered, using as a standard the level achievable by the majority of farmers and herdsmen through their own efforts. The standard should take into consideration both the actual situation in the country and the developmental plans for the future. In short, a standard which is being formulated or revised should be ambitious but achievable.

After a new standard has been drawn up, approved and implemented, it must be strictly followed. The importance of the standard should be stressed. Once a national standard, a ministry standard or an enterprise standard has been officially approved, no changes should beallowed. The technical standard approved by the State is in reality a technical regulation.

---

[1]The old 1976 Standard presented in this Appendix was formally replaced with the new National Wool Grading Standard set out in Appendix B in December 1993. However, the 1993 Standard evolved from the 1976 Standard and it will take some time before all participants in the Chinese raw wool-handling industry upgrade their practices to comply with the new Standard. Therefore, anyone seeking to understand the grading and purchasing of raw wool in China in the 1990s still needs to be familiar with the contents of this Appendix.

(II)   Specifications for the formulation of the new standard

The revision of the National Wool Grading Standard (or raw wool purchasing standard) was carried out in accordance with the instructions of the State Council and the State Planning Commission. From October 1975 to April 1976, the State Standards Bureau took responsibility for the development of a new wool purchasing standard by establishing the wool standards investigation team with the participation of the Ministry of Agriculture, the Ministry of Textile Industry and the All China Union of Supply and Marketing Cooperatives. The team conducted thorough investigations in the principal wool-producing provinces or autonomous regions, such as Qinghai, Jilin, Inner Mongolia, Xinjiang, etc; it assessed wool production and wool utilisation situations; and it identified some existing problems and possible solutions in the wool-purchasing sphere. On the basis of the investigation, the team drew up a temporary draft of a new raw wool purchasing standard based on the original standard of 1957. The draft presented the following issues for consideration.

1.   Basic requirements of a new standard
The wool purchasing standard is used by a number of different groups, each with their own requirements and different levels of technical development. That is, the standard needs to be simple enough for the majority of farmers and herdsmen to understand it so that the purchasing of raw wool is facilitated but it should also facilitate industrial utilisation. Therefore, the new standard must be useful and relevant to the farmers and herdsmen, to the purchasing units at the grass-roots level, and to the wool textile industry. At the same time, the standard must be practical, scientific and technically advanced. It should serve the needs of the domestic wool textile industry as well as satisfy the requirements of raw wool producers and purchasing agencies.

2.   The scientific basis for the formulation of a new standard
There are a variety of different sheep breeds raised throughout China. Their wool can be divided into two major categories on the basis of its utilisation by the wool textile industry: (a) homogeneous fine wool used for combing wool textile production; and (b) homogeneous semi-fine wool used for the production of wool thread.

    Fine wool or semi-fine wool which is not sufficiently homogeneous is used for carding wool textiles and for making other products. It is vitally important to have a wool purchasing standard which distinguishes between the best-quality wool and wool which is produced at the transition stage of development. This promotes the further development of sheep breeding as well as the rational utilisation of wool by the wool textile industry. On the basis of the existing sheep breeds and the different uses of wool by the wool textile industry, different standards have been set up as follows:

    Fine wool standard: This includes wool produced by purebred fine wool sheep (Merino) introduced from abroad and the purebred fine wool sheep bred domestically. Fine wool produced by improved sheep breeds which has reached the quality requirement of homogeneous wool is also included.

    Standard for improved fine wool: This refers to fine wool from crossbred sheep (obtained by crossing purebred fine wool sheep and native sheep) which has not reached the quality requirement of homogeneous fine wool.

    Semi-fine wool standard: This includes wool from exotic purebred semi-fine wool sheep, from domestic purebred semi-fine wool bred locally, and from the improved crossbred sheep which has reached the quality requirement of homogeneous semi-fine wool.

    Standard for improved semi-fine wool: This refers to the semi-fine wool produced by crossbred sheep (obtained by crossing purebred semi-fine wool sheep and native sheep) which has not reached the quality requirement of homogeneous semi-fine wool.

3. Specifications for the graded divisions within the major categories

(1) The standards for fine wool and semi-fine wool should both reach the quality requirement of homogeneous wool. In addition to the fineness specification and the homogeneity requirement, these two classes of wool can be divided further on the basis of natural wool length. The demarcation line for fine wool between Grade I and Grade II is 6cm, while that for semi-fine is 7cm.

Wool length is one of the main quantitative indicators both in animal production and the wool textile industry. The specification of wool length in the standard plays an active and promotive role in the enhancement of sheep breeding and wool production. It also provides for a policy of "high price for good-quality wool and rational utilisation of good-quality wool."

(2) The standard also provides for a special price for fine wool which exceeds 8cm in length as well as for semi-fine wool which exceeds 10cm. This specification takes account of the current situation of the sheep industry in China as well as recognising the future orientation of the sheep-breeding program. In effect, the standard provides for a "special" fine wool grade and a "special" semi-fine wool grade. This also plays an active and promotive role in current breeding work and will assist in future development by encouraging the breeding of sheep which grow longer wool.

(3) Specifications are also laid down for the lower limit of wool length. These specifications aim to discourage the old habit of shearing the sheep twice or three times a year. Such a practice results in too short a wool length and a waste of raw material resources. In order to change this custom and encourage annual shearing, the new standard provides for the enforcement of a purchasing price policy which will discourage the production of short wool.

(4) Specifications are also included for wool grease and suint. The function of wool grease and suint is to maintain a healthy growth of wool and to protect the wool from external factors which damage the wool. Therefore, low wool grease and suint content can adversely affect wool quality.

An examination of the value of wool grease and suint content by wool textile mills shows that wool which contains an appropriate amount of grease and suint for more than two-thirds of the total length of the staple yields the best-quality wool products and requires less raw wool to make a given amount of top or yarn.

A close relationship exists among the quality and quantity of wool grease and suint, sheep breeds, geographical and climatic conditions, and feeding management levels. In order to maintain excellent fine wool sheep breeds and good-quality wool products, and to allow for the actual situation of different areas, the minimum requirement of wool grease and suint for Grade I fine wool is for the wool to have an appropriate grease and suint content for at least 3cm of the staple. As for semi-fine wool, the standard only stipulates "with grease and suint" in the wool because semi-fine wool is not as dense as the fine wool, and the grease and suint content is inferior in most areas.

(5) Stipulations for the grading of improved wool. Improved wool is the wool produced by sheep which are still in the upgrading stage of crossbreeding development. On the basis of the extent of development and the morphological features of wool, improved wool is classified into two grades based on the degree of homogeneity: improved Grade I is mainly homogeneous wool and improved Grade II is mainly heterogeneous wool. There is no further classification for improved black-and-white wool, which is generally treated as simply improved black-and-white wool.

Improved wool is a kind of transition wool which has no specific types and thus should be less strict in its technical indicators. Therefore, in addition to textual specification, actual specimens could be used for reference to assist with grading. Grade quality is assessed on the morphological features of wool and the amount of withered hair and kemp in the staple.

4. Purchasing method

Wool grading: There was formerly no unified grading procedure throughout the country for purchasing wool. The old standard distinguished the wool grades according to the amount of fleece wool, with grading demarcations of 80%, 70%, 50% and 20%. Another method relied on piece wool for grading, while others relied on a combination of fleece wool and piece wool. The grading criteria varied from province to province because of the different historical practices and actual field conditions. The new revised standard contains specifications not only for the amount of fleece wool but also for the proportion by weight of loose wool or piece wool. There is no intention to impose a unified grading procedure; this would hinder the implementation of the new standard. At present, it is allowable to apply a practical grading procedure based on the shearing techniques and shearing tools used, or on fleece wool that is incomplete because of poor handling in the marketing chain. A continuous accumulation of experience can create the conditions for a transition to a national unified grading procedure. The attainment of complete fleece wool after shearing will be greatly appreciated both at home and abroad. It is good practice to strictly distinguish good-quality wool from poor-quality wool and to separate side wool, wool grown on the head, legs and tail from the fleece wool. As long as the fleece wool is intact, it is very easy and convenient to grade the wool at the purchasing stage and/or prior to its being processed at the wool textile mills.

Removal of foreign substances: The removal of foreign substances is carried out during purchasing. The purchase price of wool takes into account "shaking the fleece wool by hand to get rid of foreign substances before weighing". Therefore, the removal of foreign substances such as soil, sand and vegetative materials is called "deduction percentage". If it is impossible to shake off the foreign substances, an estimation of impurity should be given. This method which has been used for years is very primitive and scientifically irrational. The reason is that the soil and sand adhering to the wool could drop off at any time; it is impossible to shake it off once and for all. In addition, the method of shaking is not very specifically defined, which might cause disputes during purchasing, especially from those farmers who do not understand it fully and who object fairly strongly to this removal method.

Another method is to impose a lower wool-purchasing price without enforcing any deduction percentage, such as in Xinjiang Autonomous Region. In Jiangsu and Zhejiang Provinces, the wool purchasing is based on grading after washing. There is an urgent need to implement a purchasing method based on clean-wool content. In recent years, purchasing experiments have been conducted in several provinces and autonomous regions, and preliminary information has been gained. Pricing based on clean-wool content is a developmental objective. Experiments are also being planned for provinces where conditions are ripe for the introduction of such a purchasing method so that by an accumulation of experience the new purchasing method may be gradually extended.

(III)   Technical criteria for the formulation of the standard

1.   Standard for fine wool and improved fine wool

This standard is used for the purchasing of fine wool and improved fine wool produced in China.

(1) Stipulations for grading

    A. Technical requirements for fine wool

The technical requirements for fine wool are listed in Table A1.

    (a) When the quality features and grease and suint of the wool are the same as Grade I, and when the fineness is over 60 count and the length is 8cm or more, such wool can receive a quality differential ratio of 124%. This wool is packed separately. (Note: This wool is classified as "special fine wool" during purchasing although formally the grading standard does not specify "special fine wool" as a separate grade.)

(b) Fine wool of yearling sheep (i.e. at their first shearing) may have withered top of staple, cone-shaped wool tip, poor fineness or poor uniformity in staple length. It is allowed to contain foetal hair. The fineness of the wool produced by young rams is not allowed to be coarser than 58 count.

(c) When grading is in accordance with the staple length, at least 60% of the fine wool should meet this requirement.

Table A1 Technical Requirements for Fine Wool

Grade	Fineness	Length	Grease and suint	Quality features	Quality differential ratio
	(count)	(cm)	(cm)		(%)
I	60s and above	6.0–7.9	3.0 and over 3.0	All the wool should be natural white and homogeneous fine wool, uniform in fineness and length of wool staple, normal crimpness, soft feel, with elasticity even at the tip of the wool. Part of the tip of the staple can be dry or have thin wool, but without withered hair and kemp.	114
II	60s and above	4.0–5.9	Less than Grade I	Has the same quality features as Grade I, but the length and grease and suint are less than Grade I; or the length and grease are the same as Grade I but the fineness and uniformity are less than Grade I, with loose and open staple, sub-normal crimpness, and poor elasticity.	107

(d) In areas where shearing is carried out twice per year, the price should be reduced by 50% for wool which has a length of less than 4cm. The same applies both to autumn wool and to summer wool. In southwest China, the price of the wool mentioned above should be reduced by 40%. However, the actual price should be set and controlled by the provinces or the autonomous regions.

(e) The fine wool from the head, legs and tail need not be graded and should be packed separately. The price is about 40% of the price of the standard grade.

B. Technical requirements for improved fine wool

Grade I: All the wool should be natural white and basically homogeneous wool with obviously improved features. The staple consists mainly of fine down hair (true fine wool) with little coarse down hair and few heterotypical fibres. The fineness, the uniformity of wool length, crimpness, grease and suint, and apparent morphology are inferior to those of fine wool. The staple is open, with the wool tip at the top. It is allowed to contain a trace quantity of withered hair or kemp. The wool which meets the above quality requirement or the actual specimen should account for 70% or more of the fleece wool or loose wool. The quality differential ratio of this grade is 100%.

Grade II: This consists of white and heterogeneous wool with or without hair plaits. The finer staple consists of down hair, heterotypical wool and small quantities of coarse wool. There is some crisscross wool, and some withered hair and kemp. The coarser staple consists of down hair, heterotypical wool and coarse wool with more withered hair and kemp. From apparent morphology, it appears to possess the special features of improved fine wool but it also still exhibits some of the morphological features of wool from native breeds of sheep. In comparison with wool from native sheep, the proportion of down hair and heterotypical wool has increased, as also have the grease and suint. The wool which meets the above requirements and matches the appropriate specimen should account for more than 50% of the area of the fleece wool or 30% of the weight of the loose wool. The quality differential ratio of this grade is 91%.

C. It is not permissible to shear the improved fine wool sheep in autumn and summer. If such wool is offered for sale to the State, the price is to be set at 60% of the price for Grade II wool or lower. The actual price is set by the provinces or the autonomous regions.

D. The improved black-and-white wool need not be further graded at all. The whole of the improved black-and-white wool is incorporated into one grade, irrespective of the darkness of the colours or the amount. It should be packed separately and priced at 66% of the standard grade price.

E. The price for the wool shorn or pulled from raw sheepskins is set at 60 to 80% of the price for the same category and grade of wool shorn from live sheep. This kind of wool must be packed separately.

F. For each grade of fine wool and improved fine wool, the proportion of mixed grade should not exceed 10%. Mixed grade refers to the mixture between this grade and an upper grade or the lower grade. Mixing the grade by skipping a grade up or down is not permitted. If the proportion of mixed grade is found to exceed 10%, regrading is required before purchase.

G. Boxes of wool specimens of the various grades which are made available to assist with the purchasing of wool should be prepared in accordance with the textual standard for fine wool and improved fine wool. Both the textual standard and the certified specimens are equally authentic. The staple in the last row of the actual specimen should be the lower limit. (For fine wool Grade I, the last staple is the lower limit of 60 count.) The actual specimen has both the national basic standard and local imitation standard.

H. Defective wool

(a) Wool which has been branded with pitch (tar) and paint can seriously affect the quality of wool products. It is very important to persuade farmers not to use pitch or paint to mark sheep for identification purposes. Pitch wool and paint wool should be packed separately. The price is only 60% of that for the same grade. If such wool is found in the normal wool, re-grading is needed before selling. It is permissible to shear away the wool tip stained by pitch and paint before shearing. The wool whose tip has been freed of pitch and paint should be treated as normal wool.

(b) Yellow-stained wool is priced at only 65% of that for the same wool category and the same grade. The yellow-stained wool which has lost substantial wool strength should be priced on the basis of its wool strength.

(c) Double clip wool, seedy and burry wool, cockleburr-seedy wool, rigidly cotted wool, and the wool shorn or pulled from the tanned sheepskin should be priced according to the actual condition of the wool. It is up to the provinces or the autonomous regions to set the actual price based on the wool quality.

(2) Inspection method

A. Wool is packed by category and grade, and inspections should be conducted in batches.

B. Inspection quantity for each batch: In general, one bale is taken from every 20 bales for inspection. If the batch is less than 20 bales, one bale should still be taken out for inspection. Where the batch exceeds 100 bales, one bale is taken from every 50 bales, starting from the 100th bale. Where the wool is in bulk, the weight of the wool should be estimated and each 80 kg should be regarded as one pack.

C. When an inspection of wool bales is conducted, a sample should be taken from the middle after the bale is open. With fleece wool, three fleeces should be taken from each pack, while with loose wool, 5kg should be taken.

D. The sample is then divided into three equal parts. One part is used for grading according to the standard, and then the percentage of each wool grade identified is calculated. In case of disagreement, another part is used for re-inspection and the average grading is adopted. The third part is reserved for reference.

E. Fineness inspection: Comparison should be made in accordance with the lower limit of the actual specimen.

F. Length inspection: The natural length of staple is measured, i.e. the length from the base to the tip of the wool. The site chosen for such measurement is mainly the body side in combination with the area of fleece wool. The staple length of wool in bulk is measured on the basis of weight, but all the measurements should meet the stipulations.

G. Grease and suint examination: This is carried out on the area from the base to the stained-layer's tip of the staple.

H. Foreign substance inspection: At present, it is still permissible to use the current wool-purchasing method of grading the wool on the basis of clean wool, or the method of "shaking the fleece by hand to get rid of foreign substances", or the method based on raw wool.

(3) Check-and-accept, packing and labelling

A. At places where there is a fibre inspection organisation, the wool samples should be sent to that organisation for inspection, and the result should be used as the basis for setting the purchasing price. At places where no fibre inspection organisation is available, the local wool purchasing institution should be responsible for local inspection before shipping, and the final buyer should check the wool in time. If a disagreement arises after inspection, the institutions concerned should re-check as early as possible and work out a solution.

B. The wool should be packed in accordance with the place of delivery, categories and grade. The labels should be clearly written in printed characters at both ends of the bale, with dark indelible ink being used. The contents of the label should include: the delivery institution, the categories (fine wool, improved fine wool, or improved black-and-white wool), grade, colour, weight, batch number and bale number.

2. Standard for semi-fine wool and improved semi-fine wool

This standard is used for the purchasing of semi-fine wool and improved semi-fine wool.

(1) Stipulations for grading

A. Technical requirements for semi-fine wool

The technical requirements for semi-fine wool are listed in Table A2.

Table A2 Technical Requirements for Semi-Fine Wool

Grade	Fineness	Length	Grease and suint	Quality features	Quality differential ratio
	(count)	(cm)	(cm)		(%)
I	46s to 58s	7.0–9.9	With grease and suint	All the wool should be natural white and homogeneous semi-fine wool with uniform fineness and wool length. Light and large crimpness, good elasticity, nice lustre, plain wool tip or small sharp wool tip and small hair plait with the shape of plied yarn. Thick hair plait for coarse semi-fine wool, but without withered hair or kemp.	114
II	46s to 58s	4.0–6.9	Ditto	Ditto	107

(a) When the quality features, fineness, grease and suint of the wool are the same as for Grade I, and the staple length is 10cm or more, such wool can receive a quality ratio of 124%. This wool should be packed separately. (Note: This kind of wool is classified as "special semi-fine wool" during purchasing although formally the grading standard does not specify "special semi-fine wool" as a separate grade.)

(b) Purebred homogeneous wool which is stronger than 46 count wool should be graded, for the time being, according to this standard. The semi-fine wool shorn from one-year-old sheep (i.e. the first shearing) has a withered, cone-shaped wool tip. There is little uniformity in wool fineness and staple length. Foetal hair is allowed.

(c) When the semi-fine wool is graded in accordance with the staple length, 60% or more of the wool should meet the stipulations.

(d) In areas where shearing is carried out twice per year, the purchasing price for wool shorn in autumn and summer with a staple length shorter than 4cm should be reduced by up to 50%. In southwest China, the price of the same wool should be reduced by up to 40%. The actual price should be set and controlled by the local provinces and autonomous regions.

(e) The semi-fine wool shorn from the head, legs and tail should be packed separately. The price is about 40% of the price of the standard grade.

B. Technical requirements for improved semi-fine wool

Grade I: All the wool should be natural white in colour and basically homogeneous wool with obvious improved features. The staple consists mainly of coarse down hair and heterotypical wool. It is a little inferior in the uniformity of wool fineness and staple length, as well as in crimpness, grease and suint, and apparent morphology. It is allowed to contain small quantities of withered hair and kemp. The wool which meets the above requirement should account for 70% or more of the fleece wool or loose wool. The quality differential ratio of this grade is 100%.

Grade II: This consists of white and heterogeneous wool with improved features. The finer staple consists of coarse down hair, heterotypical wool and some coarse wool. There are some hair plaits at the upper part of the staple, and small quantities of withered hair and kemp. The coarser staple consists of coarse down hair, heterotypical wool and coarse wool with large quantities of withered hair and kemp. There are long hair plaits at the top of the staple. However, in comparison with the wool of native breeds, there is more down hair, and more grease and suint at the base of the staple. The wool which meets the above requirements and matches the appropriate specimen should account for more than 30% of the area of the fleeced wool or the weight of the loose wool. The quality differential ratio of this grade is 91%.

C. It is not permissible to shear the improved semi-fine wool sheep in autumn and summer. If such wool is sold, the price is to be set at 60% of Grade II wool or lower. The actual price is set by the provinces or the autonomous regions.

D. The improved black-and-white wool need not be graded at all. The whole of the improved black-and-white wool is classified into one grade, irrespective of the darkness of the colour or the amount of coloured wool. It should be packed separately and priced at 66% of the standard grade price.

E. The price for the wool shorn or pulled from raw sheepskins should be set at 60–80% of the price for the same category and grade of wool from live sheep. This kind of wool must be packed separately.

F. For each grade of semi-fine wool and improved semi-fine wool, the proportion of mixed grade should not exceed 10%. Mixed grade refers to the mixture between this grade and the upper grade or the lower grade. Mixing the grade by skipping a grade up or down is not permitted. If the proportion of mixed grade is found to exceed 10%, regrading is required before any purchase is made.

G. The actual specimen should be prepared in accordance with the textual standard for semi-fine wool and improved semi-fine wool. Both the textual standard and the actual specimen are equally authentic. The staple in the last row of the actual specimen should be the lower limit. The actual specimen has both the national basic standard and the local imitation standard.

H. Defective wool: The same technical requirements as for improved fine wool.

(2) Inspection method
The inspection method for semi-fine wool and improved semi-fine wool is the same as that for fine wool and improved fine wool.

(3) Check-and-accept, packing and labelling
The check-and-accept, packing and labelling for semi-fine wool and improved semi-fine wool are the same as those for fine wool and improved fine wool.

(IV) Glossary

(1)   Homogeneous fleece/wool: This refers to the fact that every staple in the fleece wool consists of homogeneous fibres, and the fineness and length of the fibres within the staples tend to be uniform.

(2)   Basically homogeneous fleece/wool: This refers to the fact that most of the staples in the fleece wool consist of homogeneous fibres with very few heterogeneous fibres; or the fleece wool consists mainly of fine down hair, with very little heterogeneous fibres or kemp, and few quantities of coarse down hair.

(3)   Heterogeneous fleece/wool: This refers to the fact that every staple on the fleece wool consists of more than two fibre types (mainly fine down hair and coarse down hair with uneven quantities of fine strong wool, coarse strong wool, withered hair and kemp). In most cases, there are hair plaits.

(4)   Fineness: The fineness of wool fibre is expressed in microns (μm) which are one thousandth of a millimetre.

(5)   Count: This is a method used to distinguish the fineness of wool fibres, and is also called the quality number. It is different from the quality count used in the wool textile industry. The quality number can only reflect the quality count that could be spun, and does not readily correspond with the quality count used in the wool textile industry.

(6)   Uniformity: This refers to the differential extent of the fibre fineness and staple length, or the differential extent of the fibre fineness and fibre length in different areas of the fleece wool.

(7)   Fine down hair: This refers to the wool fibre which has the most crimpness with uniform fineness. The fineness is under 30.0μm. The cross-section of the fibre is round and near-round in shape. The down hair at the middle and bottom of the heterogeneous fleece is also called fine down hair.

(8)   Coarse down hair: This refers to the wool fibre which is coarser than fine down hair but finer than strong wool. It has uniform and mostly uniform fibres. Its fineness is between 30.0 and 52.5μm.

(9)   Fine strong wool: This refers to the wool fibre which is straight and long with no, or very little, crimpness and with good elasticity and strength. The fineness is between 52.5 and 75.0μm.

(10) Coarse strong wool: This refers to the wool fibre which is straight and long with no crimpness at all. There are hair plaits at the tip of the staple. The fineness is usually over 75.0μm.

(11) Heterotypical fleece: This is the intermediate type whose fineness lies between hair and down hair. There are both cardiac layers and non-cardiac layers in the one fibre.

(12) Hair: This term is used to refer to both fine strong wool and coarse strong wool. Generally speaking, this kind of wool has no crimpness, and has strong and straight fibres. The mean fibre diameter is usually over 52.5μm.

(13) Withered wool: This is a kind of strong wool, white in colour, finer than kemp, has little strength, and is easy to break.

(14) Kemp: This is a kind of short and straight fibre which is coarse and easy to break, is withered white in colour, and has no strength. The fibre is flat in shape when observed by the naked eye, and it has no textile value.

(15) Black-and-white wool: All the coloured wool shorn from coarse wool sheep or improved crossbred sheep, and the wool shorn from white wool sheep which contains mixed colour wool or wool with single-coloured fibres, belong to this category.

(16) Crisscross wool: Wool fibres which have crisscrossed owing to uneven growth belong to this category. In general, this kind of wool contains heterotypical and coarse fibres. This wool category is usually produced by the native sheep or improved sheep of poor breeding.

(17) Elasticity: This refers to the ability of the wool to recover its original shape and size after the removal of outside mechanical force.

(18) Double clip wool: This refers to the short wool re-shorn from the sheep after the first shearing during which the stubble was left too long.

(19) Hair plait: This refers to the coarse wool fibres within heterogeneous fleeces, which are longer than the staple and form plaits at the top.

(20) Wool tip: In general, the upper part of the staple is called the wool tip. There are two categories: plain wool tip and sharp wool tip. The plain wool tip usually has a uniform shape from the top to the bottom of the staple.

(21) Yellow-stained wool: This refers to the wool which has turned yellow in more than half of the staple length. There are two categories of yellow-stained wool: water yellow-stained wool caused by the leaching of water, and pen yellow-stained wool caused by contamination by urine and dung in the pen.

(22) Fleece wool: This refers to the wool shorn in spring. The staples are joined together and form a fundamentally intact fleece. Owing to the physical condition of the sheep and poor management techniques, some cases of broken fleeces and short staples might arise. Nevertheless, such wool should be purchased according to all stipulations laid down for fleece wool.

The *(New)* National Wool Grading Standard of the People's Republic of China

(This wool standard was issued by the National Supervisory Bureau of Technologies on 28 April 1993 and implemented from 1 December 1993)

1. The scope

This standard stipulates the technical criteria for wool grading and classifying, inspection methods, and rules for inspection, packing, labelling, storage and transportation of fine, semi-fine and improved wool. The standard is applied to the production, trading, utilisation and control of fine, semi-fine and improved wool.

2. Recommended standards

GB	6976	The method for testing the natural length of wool.
GB	6977	The method for testing the grease and suint, dust and other non-wool material contents of scoured wool.
GB	6978	The method for testing the clean yield of raw wool: Oven-dry method.*
GB	10685	The method for testing the diameter of wool fibre: Image-projecting microscope method.
GB/T	14270-93	The method for testing the proportion of different types of wool in each lot (or batch).
GB/T	14271-93	The method for testing the clean yield of raw wool: Oil-press method.*

3. Glossary of terms

Fine wool:	Homogenous wool with spinning count or quality number of 60s and above with average fibre diameter ≤25.0μm.
Semi-fine wool:	Homogeneous wool with spinning count or quality number between 36s and 58s and with average fibre diameter between 25.1 and 55.0μm.
Improved wool:	Heterogeneous wool from hybrid sheep being raised during the process of breed improvement.
Fineness:	An indicator of the thickness of wool fibres, expressed as the mean fibre diameter in micron (μm) or the spinning count or quality number.
Spinning count or quality number:	An indicator of the fineness of wool fibres. Table B1 lists the values of the spinning count or quality number corresponding with particular mean fibre diameters.
Coarse wool:	Wool diameter above 52.5μm. Usually coarse wool has wool marrow and few or no crimps and is longer than fine hair.
Kemp:	The dead, white wool fibres which break easily.
Withered wool:	A type of withered and white coarse wool fibre with some tensile strength.
Sharp wool tip:	The outer or upper extremity of a staple of wool with sharp shape.
Wool plait:	Obvious cotted wool at the top of a wool staple.

---

*These two methods for obtaining estimates of the clean (scoured) yield are briefly described in Chapter 6 (Section 6.3.2).

Table B1 Wool Spinning Count or Quality Number and Corresponding Mean
Fibre Diameter

Fine wool		Semi-fine wool	
Spinning count or quality number	Mean fibre diameter	Spinning count or quality number	Mean fibre diameter
(s)	(µm)	(s)	(µm)
70	18.1–20.0	58	25.1–27.0
66	20.1–21.5	56	27.1–29.0
64	21.6–25.0	50	29.1–31.0
60	23.1–25.0	48	31.1–34.0
		46	34.1–37.0
		44	37.1–40.0
		40	40.1–43.0
		36*	43.1–55.0

*Some wool classified as having a 36s spinning count would strictly be classified as coarse wool in China since its mean fibre diameter would exceed 52.5µm.

Homogeneous fleece:	Fleece consists of only one type of wool fibre.
Heterogeneous fleece:	Fleece consists of more than two types of wool fibre.
Basically homogeneous fleece:	Fleece consists basically of one type of wool fibre with very little heterogeneous wool.
Heterotypical fibre:	Both fine and coarse wool morphology present in one wool fibre.
Skirtings:	Wool removed from the edge of the fleece, the quality of which is lower than the main part of the fleece.
Faulty wool:	A general term for pitch- or paint-stained, yellowish, dung-stained, seedy/burry, heavily cotted, scabby and tender wool.
Paint- or pitch-stained wool:	Wool contaminated with paint or pitch.
Yellowish wool:	Wool with more than half of the staple length stained with yellowish colour.
Dung-stained wool:	Wool heavily contaminated with faeces.
Seedy/burry wool:	Wool with vegetable matter contents exceeding 4% of the net wool weight.
Heavily cotted wool:	Wool which cannot be torn easily into separated fibres. The tensile strength of the wool is markedly reduced.
Scabby wool:	Wool from the scabby-infested parts of sheep.
Tender wool:	Wool with a low tensile strength and markedly thin fibres, caused by diseases or malnutrition.
Second-clip wool:	Short pieces of wool staples that result from a shearer going twice over the same area.

4.   Technical criteria

4.1   Criteria stipulated for wool classifying and grading are listed in Table B2.

4.2   The fineness and natural length of wool and the contents of grease/suint, coarse, withered and kemp wool are used as indicators for grading fine and semi-fine wool. The grade of a lot (or batch) of wool is determined by the wool possessing the lowest quality of these indicators. For grading improved wool, the staple length and contents of coarse, withered and kemp wool are used as indicators. The grade of improved wool is determined by the wool possessing the lowest quality of these two indicators. Morphological characteristics listed in the Table B2 serve as references for grading.

4.3   Fleece wool after the removal of edge wool (skirtings) is graded on the basis of natural wool length. Special grade wool must contain ≥70% weight for weight (w/w) of wool with the required staple length as listed in Table B2 and the rest with no shorter than 60mm for fine wool and no shorter than 70mm for semi-fine wool. The Grade I of fine wool must contain ≥70% (w/w) of wool with the required staple length and the rest with no shorter than 40mm (wool with the staple length between 40 and 50mm should not be ≥10%). The Grade I of semi-fine wool must contain ≥70% (w/w) of wool with the required staple length and the rest with no shorter than 60mm (wool with the staple length between 60 and 70mm should not be ≥10%). The Grade II wool must contain ≥80% (w/w) of wool with the required staple length and the rest with no shorter than 30mm for fine wool and no shorter than 50mm for semi-fine wool.

4.4   Fineness of wool fibre is used as an indicator for classification of fine and semi-fine wool. The average fineness must meet this stipulation.

4.5   For grease/suint indicator, at least 70% of the wool in the lot (or batch) must meet this stipulation.

4.6   Improved wool which contains ≥5% of coarse/withered/kemp wool or staple length of ≤40mm and possesses improved wool morphology is to be treated as out-grade and baled separately.

4.7   Improved black/white wool (black as main colour) should be baled separately regardless of the colour intensity and grades.

4.8   White/black wool (white as main colour) should be baled separately.

4.9   Skirtings or edge wool is graded on the basis of the fineness, length, grease/suint content and morphological characteristics of staples and baled separately.

4.10   Wool from head, legs and tail or other faulty wool with some commercial value should be baled separately and cannot be put into the graded wool.

4.11   Pitch- or paint-contaminated wool should be taken out and is not allowed to be put into any types of wool.

4.12   The special grade and Grade I of fine and semi-fine wool should not possess tender wool with a weak point in the middle of the wool fibre.

5.   Inspection methods

5.1   *Determining the grade and quality*

5.1.1 Sampling size
One bale is taken for inspection from every 20 bales (or from batches of wool consisting of less than 20 bales). If the number of bales exceeds 100, an additional bale is sampled from every increment of 50 bales (or part thereof).

5.1.2 The quantity of specimens
Specimens are randomly taken for inspection from the top, middle and bottom of wool bales. The total amount of specimens should not be less than 20kg from each batch or lot and not less than 3kg from each bale sampled.

Table B2 Criteria for Classifying and Grading Wool

Wool type	Grade	Fineness¹ (µm)	Natural length (mm)	Grease/suint (%)	Morphological characteristics
Fine	S*	18.1–20.0 (70s) 20.1–21.5 (66s) 21.6–23.0 (64s) 23.1–25.0 (60s)	≥75 ≥75 ≥80 ≥80	≥50** ≥50 ≥50 ≥50	No presence of coarse/withered/kemp wool. All homogeneous and naturally white wool with similar fineness and length and normal crimps. Possible presence of small sharp wool tips on some staples.
	I	18.1–21.5 (66–70s) 21.6–25.0 (60–64s)	≥60 ≥60	≥50 ≥50	The same as for special grade but with the possible presence of withered wool at the top of some staples.
	II	≤25.0 (≥60s)	≥40	+***	No presence of coarse/withered/kemp wool. All homogeneous and naturally white wool with slight difference in fineness of wool fibres. Loose structure of staples.
Semi-fine	S	25.1–29.0 (56–58s) 29.1–37.0 (46–50s) 37.1–55.0 (36–44s)	≥90 ≥100 ≥120	+ + +	No presence of coarse/withered/kemp wool. All homogeneous and naturally white wool with similar fineness and length, and big and shallow crimps. Glossy staples with plain or small sharp tips or with some small wool plaits at the wool tips.
	I	25.1–29.0 (56–58s) 29.1–37.0 (46–50s) 37.1–55.0 (36–44s)	≥80 ≥90 ≥100	+ + +	The same as the special grade of semi-fine wool.
	II	≤55.0 (≥36s)	≥60	+	No presence of coarse/withered/kemp wool. All homogeneous and naturally white wool.
Improved	I		≥60		Presence of ≤1.5% coarse/withered/kemp wool. All naturally white and basically homogeneous improved wool. Staples consist of fine hair and heterotypical wool. The homogeneity, crimps, grease/suint content and morphological characters are inferior to fine and semi-fine wool. Presence of small or medium-sized wool plaits in staple.
	II		≥40		Presence of ≤5.0% coarse/withered and heterogeneous improved wool. All naturally white and heterogeneous improved wool. Staples consist of more than two types of wool fibre with big or very shallow crimps. Presence of small or medium-sized wool plaits and grease/suint in staples.

¹ The corresponding spinning count or quality number is shown in brackets.

* Special grade.

** Percentage of the total length of staple wool containing grease and suint.

*** The staple must contain some grease and suint.

5.1.3 Inspection

Specimens are divided into three aliquots and one aliquot is taken for inspection. In the case of disagreement, another aliquot is taken for re-inspection. The average values of the results from two inspections are used to determine the grade of the lot. The third aliquot of the specimen wool is stored for reference.

5.1.4 Inspection of fineness

Specimens are examined according to the stipulated wool fineness standard. In the case of disagreement, the method described in GB 10685 will be used for further examination.

5.1.5 Inspection of the natural length of staples

Inspection is carried out according to the method described in GB 6976.

5.1.6 Inspection of wool grease and suint

Sampling is conducted according to the method described in GB 6977. The length of grease and suint-stained staples is measured and the value is expressed as the ratio of the stained length to the natural length of the staple.

5.1.7 Inspection of the contents of coarse, withered and kemp wool

Inspection is carried out according to the methods described in GB/T 14270-93.

5.2 *Official weighing and inspection*

5.2.1 Sampling

The gross weight of each bale (and of each batch or lot of bales) is determined by public weighing. At the same time, specimens are taken to determine the clean yield of wool in the batch/lot. These samples are collected from the bales using a drilling-hole method. The rate of sampling is listed in Table B3. The total weight of specimens should not be less than 1.2kg.

Table B3 The Sampling Size for Determining Clean Yield

Total number of bales in the batch/lot	25	50	75	100	150	200	300	400	500
Number of bales sampled	25	33	37	39	42	43	46	48	50

5.2.2 Inspection

(1) The weight of every bale must be measured and recorded. A balance with a maximum weighing capacity of 500kg and graduations of 0.5kg is used.
(2) Specimens are taken by a drilling-hole method (GB/T 14271-93) immediately after measurement of bale weight.
(3) The net wool rate (estimated clean yield) is determined according to the method described in GB/T 14271-93. If re-examination is required, the inspection method in GB 6978 is used.
(4) The formula for determining the amount of clean wool is as follows:

$$W = WN \times W'Y$$

where W = the amount of clean wool (kg) in the batch (or lot),
WN = the net weight (kg) of raw wool in the batch,
$W'Y$ = the clean-wool rate (or clean yield) percentage.

The official rate of moisture regain assumed is 16% for fine and semi-fine wool and 15% for improved wool.

(5) The official amount of clean wool in the batch is not adjusted if the vegetable matter content of the clean wool is ≤2% and is reduced by 1% for the presence of every 1.5% of vegetable matter content between >2% and ≤4%. If the clean wool contains >4% of vegetable matter, the lot is regarded as seedy/burry wool. The vegetable matter contents of clean wool is examined according to the method described in GB 6977.

(6) The professional wool inspection organisation should put a clear label on all bales in the batch (lot) which provides detailed results of the inspection.

(7) It should be encouraged to objectively measure the clean yield whenever the instruments are available. Visual examination is allowed if instruments for objective measurement are not available. However, objective measurement should be used often to correct the values obtained by the visual examination.

6. The inspection rules

6.1 Inspection should be carried out by the batch (or lot).

6.2 When the amount to be traded is small, all the wool being sold should be examined.

6.3 The bales should be grouped into batches or lots containing the same type of wool prior to inspection.

6.4 When a large amount of raw wool is being traded (e.g. >2,000kg), a notary public inspection system is used for determination of the grade and quality and official weight of the wool. A professional wool inspection organisation performs the examination before sale and produces an inspection certificate. The amount to be paid for the wool is based on the officially determined amount of clean wool in the batch/lot.

6.5 In the case of disagreement on the inspection results from either of the two parties, the seller and the buyer, an application for re-inspection should be sent to the wool inspection organisation within 15 days after receiving the inspection certificate. Re-inspection on the reserved specimens should be performed within 15 days after receiving the application. If the disagreement is not resolved by the second inspection, a second application for re-inspection should be lodged with a higher level of a professional wool inspection organisation in the areas where each party is located within 15 days after receiving the re-inspection certificate. For the third inspection new specimens are taken according to the stipulation in this standard. The third inspection is final and should be performed within 15 days after receiving the application.

7. Packing, labelling, storage and transportation

7.1 Packing
The sorted wool should be separately packed into tightly pressed bales on the basis of the wool category and grade and production areas. The packing should guarantee the retention of the basic morphology of the batch/lot. The appearance of wool bales should be neat and uniform. The weight of each bale should be kept in a range of 80–100kg. Linen or other durable materials should be used as packing materials. Bales are tied with a metal wire or a string at least five times around.

7.2 Labelling
A label board with a deep-coloured and permanent label is put on one end of each bale in the lot. The label contains the information about the wool production area, type, length, fineness, batch number, bale number, bale weight, the supplier's organisation and baling date, (month/year).

7.3 Storage and transportation
The labelled bales are stored in separate piles on basis of batch number and the type of wool. The storage of bales should guarantee the quality of wool. Bales with different batch number, production area and types of wool should be kept separate during transportation.

Notes

This standard was proposed by the National Textile Fibre Inspection Bureau, Ministry of Agriculture, Ministry of Commerce, Ministry of Textile Industry and National Pricing Bureau. The standard was drafted by the Department of Wool, Linen and Silkworm Cocoons, and by the National Fibre Inspection Bureau.

Appendix C

The Industrial Wool-Sorts Standard

(Issued by the Ministry of Textile Industry in 1979 and since implemented in
textile plants throughout China)

In recent years, the wool textile industry in China has continuously developed, the structure of
the domestic wool resources has greatly changed, and the output of fine wool and semi-fine
wool and their improved wool has increased year by year. Because of these advances, the
Industrial Wool-Sorts Standard issued in 1960 was far from meeting production requirements.
In 1969, the Ministry of Textile Industry charged the textile departments and woollen mills in
Beijing, Shanghai and Tianjin with the revision of the old standard. On the basis of wool
resources and quality, and the experience accumulated by the woo mills throughout China, a
revised standard was drawn up and implemented in some key wool mills on a trial basis, so
that its feasibility might be examined. In 1975, in order to review the implementation of the
revised standard, the Ministry of Textile Industry entrusted experts from Beijing, Tianjin,
Shanghai, Shaanxi, Xinjiang, Inner Mongolia, Liaoning and elsewhere with the establishment of
a coordination team to revise and amend the revised standard, with the objective of establishing
a ministry-approved wool-sorts standard. An industrial wool-sorts standard was finally issued in
1979 and has since been implemented in every woollen mill throughout the country.

(I) The aim and significance of the formulation of the Industrial Wool-Sorts Standard

The wool shipped from the producing areas or from the hand-and-accept points varies greatly in
terms of both sort-division and quality, and therefore cannot be used directly in processing.
Wool must be re-sorted according to industry requirements and must strictly comply with the
Industrial Wool-Sorts Standard. Hence the industrial wool-sort becomes a working procedure
which cannot be omitted by the woollen mills before the preliminary processing of raw wool.

The aim of establishing an industrial wool-sorts standard is to make it possible for each
type of wool to be used optimally in the wool textile industry. All the wool in each sort should
be of exactly the same type as defined in the standard for the sort in question and should have
the same technological capactiy in regard to its use in textile manufacturing. In general, wool is
categorised in accordance with the wool category, or wool-producing area, or sheep breed. Each
category should have its own technical indicators and requirements. There should be different
features and different criteria in the standard for different categories, as this is essential for
achieving a rational use of wool as a raw material.

(II) The principles for the formulation of an Industrial Wool-Sorts Standard

The formulation of an industrial wool-sorts standard must be based on the production level of
China's wool textile industry; take into consideration the existing demand for China's woollen
products and the quality of domestically-produced wool; and allow for the sorting procedures in
different woollen mills throughout China. The aim of wool sorting is to utilize the domestic
wool resources fully and rationally, enabling good-quality wool to be used for manufacturing
good-quality woollen products. This will generate a steady increase in the quality of woollen
products and a rational development of their variety. In addition, this can promote the work of
sheep improvement and breeding, and enhance animal production. At the same time, the

Industrial Wool-Sorts Standard should not only be linked with the Wool Grading Standard used for purchasing raw wool but also guarantee a smooth implementation of the Wool Top Standard; it should be adaptable not only to the requirements of combing wool but also to the characteristics of carding wool.

(III) Consideration of the content of the Wool-Sorts Standard on the basis of actual situations

The Industrial Wool-Sorts Standard of China is approved by the responsible department (the Ministry of Textile Industry) for nationwide implementation as a ministry-approved standard. This standard has been formulated on the basis of China's wool resources and the experience of wool utilisation over many years in combination with the extent of sheep improvement. The standard divides wool into two categories: homogeneous wool and heterogeneous wool. The wool which reaches the standard of homogeneous wool will be sorted by quality count. The heterogeneous wool will be further sorted by the content of coarse-cavity wool with cardiac layers and the quality features of the staple. On the other hand, wool could be broken down on the basis of sheep breeds into two major categories: fine wool and semi-fine wool. At present, the sorting standard for fine wool has been approved and issued for implementation throughout the country while that for semi-fine wool is still being formulated and drafted.

(IV) The technical criteria for the formulation of the Industrial Wool-Sorts Standard

1. *Sort division*: The standard stipulates that there are four sort divisions in the count-wool while there are five sort divisions in the sort-wool. However, in view of the fact that different regions use different wool quality counts and different sorting schemes, it is suggested that the number of sorts could be allowed to vary locally, provided the quality of each sort meets the stipulations laid down in the standard.

2. *Average fineness*: On the basis of the analysis of historical data and the results of experiment, there is a rule of fineness variance between the wool sorted by quality count and the measured fibre fineness after the combing process. Generally speaking, the combed wool is about 0.3 to 0.8μm coarser. In order to have consistency between the quality count of wool and the quality count of wool top, there will be no revisions in the fineness indicator for the time being. However, in wool sorting, it is suggested that wool up to 0.5μm finer than the standard should be taken into consideration.

3. *Fineness discreteness*: Domestically-produced fine wool should have uniformity of fineness and meet the requirements of worsted fabrics in terms of both quality and technology. However, in Inner Mongolia and north east China, sheep breeding is still at the transition stage, and some of the homogeneous fleece has poor uniformity of fineness. The deterioration in quality is caused by many factors during shearing, baling, storage, etc. It is therefore suggested that the fineness discreteness should not be applied too strictly, and the extent of fineness discreteness for domestically-produced wool could be greater than that for imported wool.

4. *Length*: Staple length is one of the important technical indicators which can show the physical properties of wool. It has a close relationship with the ratio of making top, the ratio of broken fine yarn, the ratio of making yarn, and the uniformity of yarn.

There are certain requirements for the staple length used for the production of worsted fabrics and fine thread. It is essential to sort the wool carefully so as to fully utilise the domestic wool resources by using good-quality wool for the manufacture of high-quality products. For this reason, staple length is classified into four sorts on the basis of use, as well as being linked to the wool purchasing standard so as to meet the demands of production.

5. The content of coarse-cavity wool is an important inspection indicator in the sorting standard. In addition, there are strict control stipulations in each sort. Withered hair and kemp

belong to one type of coarse-cavity wool. In general, homogeneous fine wool and semi-fine wool are not allowed to contain coarse-cavity wool. Therefore, as stipulated in the standard, sorts of 70 count and 66 count are not allowed to contain withered hair or kemp.

In addition, there are specific inspection indicators and requirements for wool grease and suint, detrimental substances, and defective wool in the Industrial Wool-Sorts Standard. These items have a substantial effect on the wool textile industry and product quality. In order to guarantee smooth product processing, and to standardise and improve product quality, it is suggested that woollen mills strengthen the wool-sorting procedure in respect of the raw wool entering the mill, and formulate measures to make effective checks on the sorting of the raw material, in the spirit of "putting more effort into sorting so as to effect a tenfold saving in other procedures along the production line".

(V)   The Industrial Wool-Sorts Standard approved and issued by the Ministry of Textile Industry

The industrial wool sorting standard for domestically-produced fine wool and improved fine wool is as follows. This standard contains unified stipulations for the identification and the hand-and-accept inspection of the industrial sorting quality of domestic fine wool and improved fine wool within the national wool textile industry.

1. Stipulations for sorting

(1) The industrial sorting of wool can be broken down into two categories according to the physical indicators and apparent morphology: count wool and sort wool.

Count wool: This is homogeneous wool which can be classified into 70, 66, 64 and 60 counts on the basis of fineness.

Sort wool: This includes both wool which is basically homogeneous and heterogeneous wool which can be classified into Sort I, II, III, IV(A), IV(B) and V on the basis of the content of coarse-cavity wool.

(2) Physical indicators of count wool (Table C1) and sort wool (Table C2).

Table C1 Count Wool

Count	Average fineness	Fineness discreteness	Rate of coarse-cavity wool	Proportion of staple length which contains grease and suint	Classification staple length
	(μm)	(%)	(%)		(cm)
70	18.1–20.0	< 24	< 0.05	> two-thirds	I > 8.0
66	20.1–21.5	< 25	< 0.10	> two-thirds	II > 6.0
64	21.6–23.0	< 27	< 0.20	> one-half	III > 5.0
60	23.1–25.0	< 29	< 0.30	> one-half	IV > 4.0

Note: The rate of coarse-cavity wool in count wool is that guaranteed under the conditions for enterprises, but no withered hair or kemp is allowed in the 70 and 66 counts. Fineness discreteness and wool grease serve as reference for the enterprises. Staple length is regarded as the classifying criterion. The proportion of wool which is lower than the lower limit of the staple length of this class should not exceed 15%, of which the proportion which is lower than the lower limit of the staple length of the next lower class should not exceed 5%. Short wool which is less than 4cm in staple length will be treated by the enterprises themselves. As for the fineness of commercially-cleaned wool used as count wool, the count should be 0.5μm finer than the stipulations laid down in each count of the standard.

Table C2 Sort Wool

Sort	Average fineness	Rate of coarse-cavity wool
	(μm)	(%)
I	< 24.0	< 1.0
II	< 25.0	< 2.0
III	< 26.0	< 3.5
IV(A)	< 28.0	< 5.0
IV(B)	< 30.0	< 7.0
V	> 30.0	> 7.0

Note: The average fineness of sort wool serves as reference only.

(3) Requirements for apparent morphology

70 count: The wool consists of fine down hair. The staple structure is neatly closed, and has an even top, good grease and suint, good lustre, obvious and uniform crimpness, and good uniformity of fineness.

66 count: Ditto.

64 count: The wool consists of fine down hair with trace quantities of coarse down hair. The staple structure is reasonably closed, and has a generally even top, fair grease and suint, fair lustre, obvious crimpness, and fair uniformity of fineness.

60 count: The wool consists of fine down hair with small quantities of coarse down hair. The staple structure is somewhat loose, and has a small wool tip, grease and suint, lustre, large crimpness, and somewhat uniform fineness.

Sort I: The wool is basically homogeneous, and consists of fine down hair with small quantities of coarse down hair or small quantities of heterotypical wool. It also contains trace quantities of coarse wool and cavity wool with large or unobvious crimpness.

Sort II: The wool consists of down hair and heterotypical wool with small quantities of coarse wool and cavity wool, or sometimes with small hair plaits or unobvious crimpness.

Sort III: The wool consists of down hair and heterotypical wool with a little more coarse wool and more cavity wool, and with thin and long hair plaits.

Sort IV(A): The wool consists of down hair and heterotypical wool with more coarse wool and obvious cavity wool, and sometimes thick and long hair plaits.

Sort IV(B): The wool consists of down hair and heterotypical wool with much more coarse wool and very obvious cavity wool, and sometimes thick and long hair plaits.

Sort V: The wool which falls below the quality requirements of Sort IV(B) but which does not belong to the defective wool category is put into this grade.

Wool which does not possess the obvious characteristics of improved wool in apparent morphology will be treated as the wool produced by native sheep.

(4) Wool sorted by the industrial wool sorting standard is not allowed to contain pitch, paint, flax fibre, chemical fibre, cotton or other foreign substances. This requirement should be established as an in-plant examination regulation.

All the defective wool such as yellow-stained wool, cotted wool, dirty-clot wool, skin-piece wool, seedy and burry wool, scabby wool, pitch wool, paint wool, etc. should be removed from the fleece for separate treatment.

(5) Coloured black-and-white wool can be sorted in accordance with the stipulations laid down in the standard.

(6) Actual specimens should be prepared in accordance with the textual standard. Both the textual standard and the actual specimen are equally authentic. The actual specimen standard has both the national basic standard (the original) and local imitation standard (the duplicate). The local imitation standard is used for the raw wool sorting on the basis of quality count and

grade. The staple in the last row of the actual specimen should be the lower limit for each count or each grade.

(7) Inspection methods should be in accordance with the "Inspection Methods for the Industrial Sorted Wool, Cleaned Wool and Wool Top of Domestic Wool".

2. Stipulations for packing and check-and-accept

(1) The packing of industrial sorted wool must guarantee the quality of the wool and facilitate transportation, handling and storage. The sorted wool must be packed in bales made of complete and smooth-surface material so as to exclude flax fibre, other kinds of fibre and other foreign substances. In the case of a contract agreement between the supplier and the receiver, the stipulations laid down in the contract should be followed.

The production mill should print on the wool bales or the labels the name of the article, sort and class, batch number, bale number, weight, place of origin, name of the mill and date of production. The receiver has the right to refuse any unclearly labelled wool bales.

(2) Check-and-accept should be handled in batches. Batches of sorted wool in each shipment should show the same place of origin, the same name of the article, the same sort, and the same contract.

(3) In the case of a dispute between the supplier and the receiver in terms of the quality of the sorted wool, it is permissible to request a re-inspection conducted by the local fibre inspection organization or any institution assigned by a higher authority. During the re-inspection, the sample should be taken in the presence of both parties. The result of the re-inspection will be final. Any fee arising from the re-inspection will be taken care of by the party which is at fault.

(4) In case of deterioration of the wool quality caused by poor handling and improper transportation and storage, it is essential to investigate and ascertain the cause. The party which is at fault should be responsible for all financial losses.

(VI) Glossary

See the Glossary at the end of Appendix A.

The *(Old)* Top Standard of Domestic Fine Wool, Improved Fine Wool and Native Wool

(Issued by the Ministry of Textile Industry and implemented in the textile plants throughout China)

(I) The top standard for domestic fine wool and improved fine wool

This standard has been formulated as a uniform specification for the identification and hand-and-accept inspection of the quality of top made from domestic fine wool and improved fine wool.

1. Technical criteria

(1) Domestic fine wool and improved fine wool are classified into two main categories: count-wool top and improved sort-wool top.

Count-wool top is further classified into 70, 66, 64 and 60 counts.

Improved sort-wool top is further classified into Sort I, Sort II, Sort III, Sort IV(A), Sort IV(B), and Sort V.

(2) Stipulations for the official rates of moisture regain and lanolin of domestic top:
A. Dry top: The official rate of moisture regain is 18.25%, while that of lanolin is 0.634%.
B. Oil top: The official rate of moisture regain is 19%, while that of lanolin is 3.5%.
C. Un-rewashed top: The official rate of moisture regain and of lanolin may be judged in accordance with the standards laid down for the dry top.
D. Packed top: The rate of moisture regain should not exceed 20%.

(3) The top should meet the stipulations laid down in Tables D1(a) and (b).

2. Stipulations for the demarcation of count, sort and class

(1) The quality count of count-wool top is determined in accordance with the average fineness of the wool. When the average fineness of the top falls outside the range specified in the regulations, the top should be sorted up or down on the basis of the actual quality count of the top.

(2) The quality sort of improved sort-wool top is determined in accordance with the coarse-cavity wool content. When such content falls outside the range specified in the regulations, the top should be sorted up or down on the basis of the actual coarse-cavity wool content.

(3) The quality class of top should be determined in accordance with the physical indicators (weighted average length, rate of short staple less than 30mm, rate of weight non-uniformity, and rate of coarse-cavity wool of count-wool top), as well as the apparent defects (wool grain, grass debris and wool piece). The top should be assigned to a class in accordance with the lowest class of that batch. Quality class is divided into Class I and Class II, while the top which is inferior to Class II should be classified as an unqualified product which, in principle, should not be shipped out of the mill.

(4) The weight per unit length of top and the official weight deviation of top are the guaranteed pre-conditions of the production mill. The unqualified top should be treated as an unqualified product.

(5) The fineness discreteness and length discreteness of count-wool top, the average fineness of sort-wool top, and the content of flax fibre and other fibre in the top are regarded as reference indicators. The colour wool of count-wool top and Sort I top is regarded as an inspection item.

(6) Each quality item and the inspected item should conform to the FJ 418-81 standard, which is entitled the "Inspection Method for Industrial Sorted Wool, Cleaned Wool and Top of Domestic Wool".

3. Stipulations for packing and check-and-accept

(1) The packing of top should be such as to maintain quality and facilitate shipping and storage.

(2) Each top ball should be packed in thick and strong kraft paper or plastic bags. Smooth material should be used to bale the top to prevent flax fibre and other kinds of fibre from entering it. In contract cases, the baling of top should strictly follow the regulations laid down in the contract.

(3) The producing mill should clearly print on the label the product name, sort, batch number, specification, bale number, weight, manufacturer, and date of production. The receiver has the right to refuse acceptance of the top if no label is attached.

(4) The export and inspection of top are carried out in batches. Each batch contains top with the same quality, and is produced by using the same technology, the same raw materials and the same formulation.

(5) Each batch of top should be inspected for its quality and official weight in accordance with the stipulations laid down in this standard. The calculation formula of the official weight is as follows:

$$\text{Official weight} = \frac{\text{AW} (1 - \text{PRMR})}{1 + \text{ARMR}} \cdot [1 + (\text{PRL} - \text{ARL})]$$

Where:
$\quad$ AW $\quad=\quad$ actual weight
$\quad$ PRMR $\quad=\quad$ official rate of moisture regain
$\quad$ ARMR $\quad=\quad$ actual rate of moisture regain
$\quad$ PRL $\quad=\quad$ official rate of lanolin
$\quad$ ARL $\quad=\quad$ actual rate of lanolin.

(6) In the inspection and acceptance of top according to the official weight, the top weight should not be changed if the weight deviation does not exceed ± 1%. However, if the weight deviation exceeds ± 1%, the top weight should be changed.

(7) In case of a disagreement about part or the whole of the inspection results by the supplier or the receiver, a re-inspection by either party is allowable. During the re-inspection, sampling may be jointly done by both parties. It is also allowable to request the local fibre inspection institution, or any institution assigned by an authority, to carry out the re-inspection. The results of the re-inspection are final, and the cost of the service should be taken care of by the responsible party.

(II) The top standard for native wool

This standard has been formulated as a uniform specification for the identification and hand-and-accept inspection of the quality of top made from native wool.

1. Technical criteria

(1) Top made from native wool is classified into three categories: Sort III, Sort IV and Xining Wool, of which Sort III and Sort IV are further classified into two sub-sorts: A and B.

(2) The official rates of moisture regain and lanolin of native wool top are stipulated as follows:

A. Dry top: The official rate of moisture regain is 18.25%, while that of lanolin is 0.634%.

B. Oil top: The official rate of moisture regain is 19%, while that of lanolin is 3.5%.

C. Un-rewashed top: The official rate of moisture regain and of lanolin may be judged in accordance with the standards laid down for dry top.

D. Packed top: The rate of moisture regain should not exceed 20%.

(3) The top should meet the stipulations laid down in Table D2.

2. Stipulations for the demarcation of count, sort and class

(1) The quality sort of native wool top is determined in accordance with the coarse-cavity wool content. When such content falls outside the range specified in the regulations, the top should be sorted up or down on the basis of the actual coarse-cavity wool content.

(2) The quality class of native wool top is determined in accordance with the physical indicators (weighted staple length, rate of short staple less than 30mm, rate of weight non-uniformity), as well as with the apparent defects. The top should be assigned to a class in accordance with the lowest class of that batch. Quality class is classified into Class I and Class II, while the top which is inferior to Class II should be classified as an unqualified product, which, in principle, should not be shipped out of the mill.

(3) The weight per unit length of top and the official weight deviation of top are the guaranteed pre-conditions of the production mill. The unqualified top should be treated as unqualified product.

(4) The average wool fineness of the top is regarded as a reference indicator.

(5) Each quality item and the inspected item should be examined in accordance with the FJ 418-81 standard, which is entitled the "Inspection Method for the Industrial Sorted Wool, Cleaned Wool and Top of Domestic Wool".

3. Stipulations for packing and check-and-accept

(1) The packing of top should be such as to maintain quality and facilitate shipping and storage.

(2) Each top ball should be packed in thick and strong kraft paper or plastic bags. Smooth material should be used to bale the top to prevent flax fibre and other kinds of fibre from entering it. In contract cases, the baling of top should strictly follow the regulations laid down in the contract.

(3) The producing mill should clearly print on the label the product name, sort, batch number, specifications, bale number, weight, manufacturer, and date of production. The receiver has the right to refuse acceptance of the top if no label is attached.

(4) The export and inspection of top are carried out in batches. Each batch contains top with the same quality, and is produced by using the same technology, the same raw material and the same formulation.

(5) Each batch of top should be inspected for its quality and official weight in accordance with the stipulations laid down in this standard. The calculation formula of the official weight is as follows:

$$\text{Official weight} = \frac{AW\,(1 - PRMR)}{1 + ARMR} \cdot [1 + (PRL - ARL)]$$

Where:
AW	=	actual weight
PRMR	=	official rate of moisture regain
ARMR	=	actual rate of moisture regain
PRL	=	official rate of lanolin
ARL	=	actual rate of lanolin.

(6) In the inspection and acceptance of top according to the official weight, the top weight should not be changed if the weight deviation does not exceed ±1%.

(7) In case of a disagreement about part or the whole of the inspection results by the supplier or the receiver, a re-inspection by either party is allowable. During the re-inspection, sampling may be jointly done by both parties. It is also allowable to request the local fibre inspection institution, or any institution assigned by an authority, to carry out the re-inspection. The results of the re-inspection are final, and the cost of the service should be taken care of by the responsible party.

Table D1(a) The Technical Criteria of Top of Fine Wool (Count-Wool Top)

Variety	Class	Physical indicators									Apparent defect			
		Average fineness	Fineness discretion	Rate of coarse-cavity wool	Weighted average length	Length discretion	Rate of short wool <30 mm	Official weight	Weight deviation	Rate of weight un-uniformity	Wool grain	Wool piece	Grass debris	Flax fibre or other fibre
		(µm)	(%)	(%)	(mm)	(%)	(%)	(g/m)	(g/m)	(%)	(no./g)	(no./g)	(no./g)	(no./g)
70	1	18.1–20.0	< 24	< 0.05	< 70.0	< 37	< 4.0	20	±1.0	< 3.0	< 4.0	< 0.30	< 0.40	< 0.10
	2	18.1–20.0	< 24	< 0.10	< 65.0	< 37	< 6.0	20	±1.0	< 4.5	< 6.0	< 0.50	< 0.60	< 0.10
66	1	20.1–21.5	< 25	< 0.10	< 70.0	< 37	< 4.0	20	±1.0	< 3.0	< 4.0	< 0.30	< 0.40	< 0.10
	2	20.1–21.5	< 25	< 0.20	< 65.0	< 37	< 6.0	20	±1.0	< 4.5	< 6.0	< 0.50	< 0.60	< 0.10
64	1	21.6–23.0	< 27	< 0.20	< 72.0	< 37	< 4.0	20	±1.0	< 3.0	< 4.0	< 0.30	< 0.40	< 0.10
	2	21.6–23.0	< 27	< 0.30	< 68.0	< 37	< 6.0	20	±1.0	< 4.5	< 6.0	< 0.50	< 0.60	< 0.10
60	1	23.1–25.0	< 29	< 0.30	< 72.0	< 37	< 4.0	20	±1.0	< 3.0	< 4.0	< 0.30	< 0.40	< 0.10
	2	23.1–25.0	< 29	< 0.40	< 68.0	< 37	< 6.0	20	±1.0	< 4.5	< 6.0	< 0.50	< 0.60	< 0.10

Notes: (1) The standard specimen of wool grain and wool piece should follow the specimen set up at the Kaifeng Conference held in 1979.
(2) The weight per unit length of top could be specified in other criteria according to the requirements of both the supplier and the receiver.
(3) The specimen for the inspection of uniformity of top could be set up by the supplier and the receiver and followed by both parties in inspection.

Table D1(b) The Technical Criteria of Top of Improved Fine Wool (Sort-Wool Top)

Variety	Class	Average fineness	Fineness discretion*	Physical indicators							Apparent defect			
				Rate of coarse-cavity wool	Weighted average length	Length discretion*	Rate of short wool <30mm	Official weight	Weight deviation	Rate of weight un-uniformity	Wool grain	Wool piece	Grass debris	Flax fibre or other fibre
		(μm)	(%)	(%)	(mm)	(%)	(%)	(g/m)	(g/m)	(%)	(no./g)	(no./g)	(no./g)	(no./g)
Sort I	1	22.0–24.0		< 1.0	< 75.0		< 5.0	20	±1.5	< 3.5	< 4.0	< 0.40	< 0.60	< 0.10
	2	22.0–24.0		< 1.0	< 70.0		< 7.0	20	±1.5	< 5.0	< 6.0	< 0.60	< 1.00	< 0.10
Sort II	1	23.0–25.0		< 2.0	< 75.0		< 5.0	20	±1.5	< 3.5	< 4.0	< 0.40	< 0.60	< 0.10
	2	23.0–25.0		< 2.0	< 70.0		< 7.0	20	±1.5	< 5.0	< 6.0	< 0.60	< 1.00	< 0.10
Sort III	1	24.0–26.0		< 3.5	< 75.0		< 5.5	20	±1.5	< 4.0	< 4.0	< 0.40	< 0.80	< 0.10
	2	24.0–26.0		< 3.5	< 70.0		< 7.5	20	±1.5	< 5.5	< 6.0	< 0.60	< 1.20	< 0.10
Sort IV(A)	1	24.0–28.0		< 5.5	< 75.0		< 5.5	20	±1.5	< 4.0	< 4.0	< 0.40	< 0.80	< 0.10
	2	24.0–28.0		< 5.5	< 70.0		< 7.5	20	±1.5	< 5.5	< 6.0	< 0.60	< 1.20	< 0.10
Sort IV(B)	1	24.0–30.0		< 7.0	< 75.0		< 5.5	20	±1.5	< 4.0	< 4.0	< 0.40	< 0.80	< 0.10
	2	24.0–30.0		< 7.0	< 70.0		< 7.5	20	±1.5	< 5.5	< 6.0	< 0.60	< 1.20	< 0.10

*Not applicable for top made from improved fine wool.

Notes: (1) The standard specimen of wool grain and wool piece should follow the specimen set up at the Kaifeng Conference held in 1979.

(2) The weight per unit length of top could be specified in other criteria according to the requirements of both the supplier and the receiver.

(3) The specimen for the inspection of uniformity of top could be set up by the supplier and the receiver and followed by both parties in inspection.

Table D2 The Technical Criteria of Native Wool Top

Variety	Class	Physical indicators							Apparent defect		
		Average fineness	Rate of coarse-cavity wool	Weighted average length	Rate of short wool <30mm	Official weight	Weight deviation	Rate of weight un-uniformity	Wool grain	Wool piece	Grass debris
		(μm)	(%)	(mm)	(%)	(g/m)	(g/m)	(%)	(no./g)	(no./g)	(no./g)
Sort III(A)	1	24.0–28.0	< 4.5	< 85.0	< 5.5	20	±1.5	< 4.0	< 4.0	< 0.40	< 0.80
	2	24.0–28.0	< 4.5	< 80.0	< 7.5	20	±1.5	< 5.5	< 6.0	< 0.60	< 1.20
Sort III(B)	1	25.0–29.0	< 6.5	< 85.0	< 5.5	20	±1.5	< 4.0	< 4.0	< 0.40	< 0.80
	2	25.0–29.0	< 6.5	< 80.0	< 7.5	20	±1.5	< 5.5	< 6.0	< 0.60	< 1.20
Sort IV(A)	1	25.0–30.0	< 7.5	< 85.0	< 5.5	20	±1.5	< 4.0	< 4.0	< 0.40	< 0.80
	2	25.0–30.0	< 7.5	< 80.0	< 7.5	20	±1.5	< 5.5	< 6.0	< 0.60	< 1.20
Sort IV(B)	1	27.0–32.0	< 9.0	< 85.0	< 5.5	20	±1.5	< 4.0	< 4.0	< 0.40	< 0.80
Knitting wool	1	28.0–34.0	< 10.0	< 85.0	< 5.5	20	±1.5	< 4.0	< 4.0	< 0.40	< 0.80
	2	28.0–34.0	< 10.0	< 80.0	< 7.5	20	±1.5	< 5.5	< 6.0	< 0.60	< 1.20

Notes: (1) The wool piece is inspected in accordance with "The Top Standard Specimen of Domestic Wool" provided by Beijing, Tianjin and Shanghai.
(2) The weight per unit length of top could be specified in other criteria according to the requirements of both the supplier and the receiver.
(3) The specimen for the inspection of uniformity of top could be set up by the supplier and the receiver and followed by both parties in inspection.

Appendix E

The *(New)* National Top Standard of Domestic and Imported Wool in China

(The manuscript for submission to the Ministry of Textile Industry
for examination and approval)

1. The scope
This standard stipulates the technical criteria, testing methods, check-and-accept, and packing of top made from domestic wool. The standard also applies to the top made from imported Australian and New Zealand wool.

2. Recommended standards
FJ 418-81   Testing methods from the national wool industry for sorted wool, washed wool and top.
GB 8170-87   Rules of value modification and approximation.
GB 9994-88   Official rate of moisture regain for textiles.

3. Technical criteria
3.1   *Stipulations for sorting*
3.1.1   The top from Australia and New Zealand is sorted on the basis of average fineness of the wool. Wool with an average fibre diameter of 25μm or less is classified as fine wool and over 25μm as semi-fine wool (see Table E1).

Table E1 Fineness Classifications used in Describing Tops

Wool category	Wool fineness (μm)							
Fine wool	19	20	21	22	23	24	25	
Semi-fine wool	26–27	28–29	30–31	32–33	34–35	35–36	36–37	38–40

3.1.2   In the case of average fineness of wool fibre being beyond the specified range in this top standard, then the wool should be sorted into other groups in accordance with its fineness.

3.2   *Stipulations for grading*
3.2.1   The quality of top is divided into two grades according to its physical properties and morphological faults. Wool top inferior to Grade II is classified as out-grade (i.e. an unqualified product). In principle, out-grade top is not allowed to be exported from the mill.
3.2.2   Guaranteed pre-conditions on weight per unit length and official weight deviation of top apply to production mills. Any unqualified top should be treated as out-grade. However, a deal between a receiver and a supplier can proceed if there is a special agreement on the contract for the unqualified top.
3.2.3   The grade of top is determined by the following physical indicators: weighted average length, the rate of the short staple below 30mm, the rate of weight non-uniformity and morphological faults (wool grain, wool piece and grass debris). The top is assigned to a grade in accordance with the lowest grade of the batch.

3.3   Technical criteria of top should meet the stipulations shown in Table E2 (p.222).

3.4   Official rate of moisture regain and official rate of lanolin for domestic and improved wool top

3.4.1 The official rate of moisture regain is 18.25% for dry top (including rewashed and un-rewashed wool) and 19% for greasy top, while the official rate of lanolin is 0.634% for dry top and 3.5% for greasy top.

3.5   The official rate of moisture regain of top at the time of packing should not exceed 20% and the official rate of lanolin of dry top should not exceed 1%.

4. Stipulations for check-and-accept

4.1   Export of top from a mill should be carried out in batches. Each batch should contain top of the same quality, produced using the same technology, the same raw material and the same formulation.

4.2   Each batch of top should be inspected for its quality and official weight in accordance with the stipulations specified in this standard. The formula for official weight is as follows:

$$\text{Official weight} = \frac{AW\ (1 - PRMR)}{1 + ARMR} \cdot [1 + (PRL - ARL)]$$

where   AW   = actual weight
        PRMR = official rate of moisture regain
        ARMR = actual rate of moisture regain
        PRL   = official rate of lanolin
        ARL   = actual rate of lanolin.

4.3   The official weight of the top should not be changed if the weight deviation does not exceed ±1% at the time of inspection. However, if the weight deviation exceeds ±1%, the official weight of the top should be changed accordingly.

4.4   In the case of a disagreement about part or the whole of the inspection by the supplier or the receiver, a re-inspection by either party is allowable. During the re-inspection, sampling may be undertaken jointly by both parties. It is also allowed to request the local fibre inspection institution of any institutions assigned by an authority to conduct the re-inspection. The results of the re-inspection are used as final specifications for the quality of the top and the cost of the service should be paid by the party responsible for the re-inspection.

5. Packing and labelling

5.1   Consideration should be given during the packing of top to the maintenance of top quality, and to facilitation of transportation and storage of the top.

5.2   Each top bale should be packed in thick and strong kraft paper or plastic bags. Smooth material should be used to bale the top to prevent packing material and other kinds of fibres from entering it. If the packing procedure is specified in the contract, the packing of top should strictly follow the specified regulations.

5.3   The production mill should clearly label on the packed top the product name, sort, batch number, specification, bale number, weight, manufacturer and the date of production.

(First Supplement)

Additional specifications

1. Fineness discretion and length discretion of wool are used as reference indicators.

2. The minimum size regarded as wool grain should be in accordance with the wool grain specimen shown in the sample photograph of wool grain and piece of national wool top; the minimum size considered as wool piece is in accordance with the specimen of two wool pieces shown in the second line at the left-top side of the sample photograph.

3. The grass debris is counted only if its length is 2mm or longer. Grass debris with a length of 15mm or longer should be recorded as two counts.

4. In the case of the quality of some top which is beyond the specified ranges in this top standard, the agreement of quality between the supplier and receiver may be reached by negotiation.

Notes

This standard was proposed by the Department of Science and Technology, Ministry of Textile Industry, the People's Republic of China. It was drafted by the Shanghai Research Institute for Wool and Jute Textile Science and Technology and was submitted to the Ministry of Textile Industry for examination and approval.

Appendix E

(Second Supplement)

(The minutes of the Consultative Meeting for the Revision of the
National Top Standard of Domestic and Imported Wool, 15 October 1989)

The Shanghai Research Institute for Wool and Jute Textile Sciences and Technology held a consultative meeting for the revision of the National Top Standard of Domestic and Imported Wool in Xitang, Wuxi on 14–16 October 1989. More than 20 delegates from Beijing, Tianjing, Jiangsu, and Shanghai attended the meeting. The aim of the meeting was to gather opinions from the delegates on the newly drafted top standard, which was based on the request for revision of the existing top standard as stated in Document 38 (1987) from the Ministry of Textile Industry. The agreement on revision of a number of important indicators for top quality has been reached after extensive discussion during the meeting. (See Table E2 below.) However, revision of some indicators still awaits further verification. The main topics discussed at the meeting were as follows.

1. Fineness as an indicator for top classing
Delegates agreed that fineness should be used for top classing instead of spinning count which was used in the previous top standard. It is considered that the top classing on the basis of wool fineness is more scientific and also in harmony with top standards in other countries.

2. Wool grain, grass debris and wool-net clearness
The previous top standard stated that any wool grain should be counted regardless of its size. Delegates considered that such a standard was impractical to manage without a limitation for the minimum size. The minimum size of wool grain is >0.5mm in IWS top standard and >0.8mm in Japanese top standard. Delegates agreed that the specimen of National Wool Grain should be used as a standard for the minimum size of wool grain. Owing to the high top quality required for making worsted fabric and knitting-wool fabric, the number of wool grain was reduced from 3.5 counts per gram of Grade I top in the original draft of the top standard to 3 counts per gram, while it remained unchanged for Grade II top.

It was also agreed that the specimen of National Wool Grass Debris should be used as a standard for the minimum size of grass debris. As the presence of grass debris greatly affects the quality of the top, the amount (number) of grass debris (0.4 count per gram of Grade I top) in the original draft of the top standard was reduced to 0.3 count per gram. The minimum length of small grass in Grade I top was changed from 1.5 to 2mm and the minimum length of long grass was changed from 10 to 15mm, which should be recorded as two counts. However, the final decision about these changes should await trails at local institutions.

It is considered that there is a close relationship between wool-net clearness and the quality of the product. Wool-net clearness is also closely related to the amount of wool grain. Thus, it is necessary to list wool-net clearness as a quality indicator. However, due to current difficulties in establishing the standard specimen, it is suggested that wool-net clearness be temporarily used as an auxiliary indicator.

3. The rate of lanolin in top
If the rate of lanolin in the finished top is too high, especially the residual lanolin, it will cause difficulties in processing products from the top. The rate of lanolin in some dry top

220

may be as high as 1.4 to 1.5%, which causes difficulty in spinning. On the basis of the present situation of production mills, it is considered that the rate of lanolin should be not higher than 1%, and that this should be listed as a reference indicator for top quality. If additional requirements are needed in some regions, then these should be resolved by consultation.

4. The rate of short wool below 30mm

It was considered that the rate of short-wool in semi-fine wool specified in the previous top standard basically meets the requirement of production mills. A reduction in the rate of short wool would improve the quality of knitting wool. However, due to the current shortage of wool, it is agreed to increase the rate of short wool from 3 to 3.5% for 26μm top and from 2 to 2.5% for 31–32μm top.

5. Length discretion and fineness discretion

Currently, there is a discrepancy in using length discretion and fineness discretion as indicators of top quality in various regions. As these properties are not used as the quality indicators in other countries and the relationship between the discretions and spinning remains to be established, delegates considered that the length discretion and fineness discretion should be listed as reference indicators only.

6. Other aspects
(1) The average length of 22μm top was reduced from 80mm in the original draft of the top standard to 75mm for Grade I, and from 75 to 70mm for Grade II.
(2) "Dry top" in Section 3.4.1 in the original draft is now specified as "Dry top (including rewashed and un-rewashed)".

The National Top Standard Revision Committee will modify the original version of the draft of the National Top Standard according to the delegates' opinions on the above six issues and data obtained from trial performance based on the original draft of the Top Standard. After these modifications, the manuscript of the National Top Standard will be submitted to the Ministry of Textile Industry for their examination and approval.

Table E2 Revised Top Grading Details as Agreed October 1989

Fineness	Grade	Physical indicator								Morphological fault		
		Range of Average fineness (μm)	Fineness discretion (%)	Average length (mm)	Length discretion (%)	Rate of short wool <30mm (%)	Official weight (g/m)	Weight deviation (g/m)	Rate of weight non-uniformity (%)	Wool grain (no./g)	Wool piece (no./g)	Grass debris (no./g)
19	1	18.6–19.5	≤22.0	≥70.0	≤37.0	≤4.2	20.0	≤±1.0	≤3.0	≤3.0	≤0.3	≤0.3
	2			≥65.0		≤6.2			≤4.0	≤5.0	≤0.5	≤0.6
20	1	19.6–20.5	≤22.0	≥70.0	≤37.0	≤4.2	20.0	≤±1.0	≤3.0	≤3.0	≤0.3	≤0.3
	2			≥65.0		≤6.2			≤4.0	≤5.0	≤0.5	≤0.6
21	1	20.6–21.5	≤23.0	≥75.0	≤37.0	≤4.2	20.0	≤±1.0	≤3.0	≤3.0	≤0.3	≤0.3
	2			≥70.0		≤6.2			≤4.0	≤5.0	≤0.5	≤0.6
22	1	21.6–22.5	≤23.0	≥75.0	≤37.0	≤4.0	20.0	≤±1.0	≤3.0	≤3.0	≤0.3	≤0.3
	2			≥70.0		≤6.0			≤4.0	≤5.0	≤0.5	≤0.6
23	1	22.6–23.5	≤24.0	≥80.0	≤37.0	≤4.0	20.0	≤±1.0	≤3.0	≤3.0	≤0.3	≤0.3
	2			≥75.0		≤6.0			≤4.0	≤5.0	≤0.5	≤0.6
24	1	23.6–24.5	≤24.0	≥80.0	≤37.0	≤4.0	20.0	≤±1.0	≤3.0	≤3.0	≤0.3	≤0.3
	2			≥75.0		≤6.0			≤4.0	≤5.0	≤0.5	≤0.6
25	1	24.5–25.5	≤24.0	≥80.0	≤38.0	≤3.5	20.0	≤±1.0	≤3.0	≤3.0	≤0.3	≤0.3
	2			≥75.0		≤5.0			≤4.0	≤5.0	≤0.5	≤0.6
26–27	1	25.6–27.5	≤25.0	≥85.0	≤38.0	≤3.5	20.0	≤±1.5	≤3.5	≤2.0	≤0.4	≤0.3
	2			≥75.0		≤5.0			≤4.5	≤3.0	≤0.6	≤0.6
28–29	1	27.6–29.5	≤26.0	≥90.0	≤39.9	≤3.5	20.0	≤±1.5	≤3.5	≤2.0	≤0.4	≤0.3
	2			≥80.0		≤5.0			≤4.5	≤3.0	≤0.6	≤0.6
30–31	1	29.6–31.5	≤26.0	≥90.0	≤39.0	≤3.5	20.0	≤±1.5	≤3.5	≤2.0	≤0.4	≤0.3
	2			≥90.0		≤5.0			≤4.5	≤3.0	≤0.6	≤0.6
32–33	1	31.6–33.5	≤26.0	≥100.0	≤41.0	≤2.5	20.0	≤±1.5	≤3.5	≤2.0	≤0.4	≤0.3
	2			≥85.0		≤4.0			≤4.5	≤3.0	≤0.6	≤0.6
34–35	1	33.6–35.5	≤26.0	≥100.0	≤41.0	≤2.5	20.0	≤±1.5	≤3.5	≤1.5	≤0.4	≤0.3
	2			≥85.0		≤4.0			≤4.5	≤2.5	≤0.6	≤0.6
36–37	1	35.6–37.5	≤26.0	≥100.0	≤41.0	≤2.5	20.0	≤±1.5	≤3.5	≤1.5	≤0.4	≤0.3
	2			≥85.0		≤4.0			≤4.5	≤2.5	≤0.6	≤0.6
38–40	1	37.6–40.5	≤26.0	≥100.0	≤41.0	≤2.5	20.0	≤±1.5	≤3.5	≤1.5	≤0.4	≤0.3
	2			≥85.0		≤4.0			≤4.5	≤2.5	≤0.6	≤0.6

Wool Quality Control Regulation for the
Inner Mongolia Autonomous Region

Clause I

In order to strengthen wool quality management and keep normal order in the wool market of the Autonomous Region, this regulation has been made according to the "Standardisation Law of the People's Republic of China" and the stipulations concerned in the National Wool Standard GB 1523-79 and GB 1524-79.

Clause II

All individuals and organisations who are engaged in wool production, purchasing, marketing and processing (simplified as producer, buyer and user in the following) in the region, must comply with this regulation.

Clause III

The Fibre Inspection Bureau of the region is authorised to act as the agent to supervise and control wool quality within the region. It is responsible for quality testing, supervision and control in the process of wool producing, purchasing, marketing and processing; for inspecting and penalising any conducts which are against the National Wool Standard in the process of wool marketing, for supervising and guiding the producers, buyers and users to comply correctly with the National Standard; and for mediating and arbitrating any disputes associated with wool quality.

Clause IV

The National Wool Standard is a compulsory standard imposed by the State. Producers, buyers and users must comply with it. Qualification examinations must be made by the Bureaus of Industry, Commerce and Administration before any business certificate is issued.

1. Producers should actively learn and master the National Standard. Wool must be classified and graded according to the standard before its delivery. The quality of the wool purchased by a contract with the State must be guaranteed in the light of the contract agreement. Upgrading and adulteration is forbidden.
2. Buyers must abide by the National Grading Standard when the wool is purchased and sold. Overgrading or undergrading against the National Grading Standard is forbidden. The wool which is purchased and sold must be classified and graded according to the regulations concerned.
3. Users must organise and accept wool according to the National Wool Standard. If the quality of the wool delivered by a wool trader is higher than the National Standard and accepted by users, the users must pay for the wool according to its quality.

Clause V

The Fibre Inspection Bureau of the Inner Mongolia Autonomous Region and the specialised inspection agencies entrusted by the Bureau are responsible for the pre-sale notary inspection of wool which is traded between industrial sectors and commercial sectors, between industrial sectors and animal husbandry sectors, or within commercial sectors, or at specialised wool markets and at wool auctions.

1. Sellers must apply for the notary inspection in time from the fibre inspection agencies. Wool traded in bulk must be sold on a clean basis. Trading without notary inspection is forbidden.

2. The result of the notary inspection provided by the fibre inspection agencies is the basis from which the price of wool is calculated by the two sides of the deal.
3. For wool which is already inspected, the sellers must guarantee the identity of the inspection certificate attached to the wool and guarantee the attachment of the certificate to the wool.
4. Inspection fees must be paid in time to the fibre inspection agencies according to the State and the region's regulations.

Clause VI

If any side to the trade has objections against the result of notary inspection provided by the fibre inspection agencies, it is entitled to apply to the Fibre Inspection Bureau of the Autonomous Region for a second inspection within seven days from the date of receiving the inspection certificate. If the trader again has objections against the result of the second test conducted by the Fibre Inspection Bureau of the region, it is entitled to apply for arbitration by the Technical Superintendence Bureau of the region or the Fibre Inspection Bureau of China. The side responsible must make compensation for any economic losses of the other side due to the irresponsibility of wool quality control judging by the result of re-testing or the arbitration. The fibre inspection agencies have to pay for the re-testing fees and arbitration fees arising from technical failure of the test.

Clause VII

If one of the following conducts is committed, a fine will be penalised by the following rules:
1. Purchasing or receiving wool without the implementation of the National Grading Standard by decreasing or increasing one level or grade. All the illegal income will be confiscated. In addition, a 10% fine of the gross value of that wool lot will be levied, and the responsible persons will also be penalised up to ¥2,000.
2. Adulterating in wool dealing activities. All the adulterated wool will be confiscated. All the illegal income will be confiscated if the wool is sold. The responsible persons will be fined up to ¥5,000.
3. Buyers classing and grading the wool purchased without following the National Standard and marketing wool with a mixed grade. The buyers must be ordered to stop any marketing activities and rectify the style of work in a given period of time. If the wool is traded without inspection, errors must be rectified and the buyers will be fined up to ¥2,000 depending upon the seriousness of the matter.

The instructions in this clause such as stoppage of marketing activities or rectification of errors in a given period of time, may be decided by the responsible administrative sectors. The other administrative penalties may be decided by the Fibre Inspection Bureau of the region and the Industrial, Commercial and Administrative Agency of the region within the terms of reference.

Clause VIII

The person concerned who has objections against the penalty decision is entitled to apply for a reconsideration to the responsible department of one level higher within 15 days from the date of pronouncing or to make a direct prosecution to the People's Court. When the given period of time expires and the person concerned fails to apply for the reconsideration or fails either to make a prosecution to the People's Court or to execute the penalty decision, the penalty decision will be executed by the People's Court at the request of the decision-making agency.

Clause IX

Compensation to others cannot be remitted from the penalties in Clause VII of this regulation. The victims are entitled to claim a compensation for any losses from the responsible persons. The Fibre Inspection Bureau of the Autonomous Region is responsible to accept and settle the dispute associated with the compensation responsibility and the amount of the compensation. The person concerned can also make a direct prosecution to the People's Court about the dispute.

Clause X

Any staff that are in charge of quality control and management in the fibre inspection agencies will be punished with disciplinary sanction by the responsible departments and where crimes are committed, the criminal responsibility will be investigated by judicial departments according to the law if the staff commit one of the following acts:
1. Cause severe economic losses due to breaking the rules in this regulation and working errors.
2. Make up or falsifying the examined data.
3. Malpractices, abuse of power, bribery.

Clause XI

All income accruing from penalties imposed will become part of the finance of the Autonomous Region. The penalty on an institution must be paid from its own money and must not be taken from production costs. The penalty on the responsible person cannot be refunded from public funds.

Clause XII

The Fibre Inspection Bureau of the Inner Mongolia Autonomous Region is responsible for the interpretation of this regulation.

Clause XIII

This regulation comes into effect from the date of issue.

Economic Commission of IMAR
Technical Superintendence Bureau of IMAR
20 April 1992

National Wool Quality Control Regulation
– Control and Management of Wool Quality –

(The document issued by the National Supervisory Bureau of Technologies, National
Economics and Trade Committee, National Planning Committee, Ministry of Agriculture,
Ministry of Textile Industries, Ministry of Commerce on 3 April 1993)

1. In order to enforce the control and management of wool quality and to maintain the order of
wool marketing and to protect the national benefit and legal rights of wool-trading parties, a
proper measure is formulated in this document on the basis of the law and stipulation of
product quality and standards and the related documents issued by the departments of the State.

2. Every person or organisation who is engaged in wool trading in the People's Republic of
China must comply with this regulation.

3. Both wool-trading parties must strictly follow the stipulation for wool classifying and
grading, wool quality control and net wool pricing laid down in the National Wool Standard. In
large-scale trading, wool pricing must be on the basis of net wool. Illegal behaviour detrimental
to wool quality, such as addition of false material into wools is prohibited.

4. Both wool-trading parties should have the related documents of National Standards and
standard specimens, must strictly comply with check-and-accept rules of wool trading and
understand the responsibility for the guarantee of wool quality.

5. When the amount of wool being traded is relatively large, trading should follow the notary
public inspection system. Wool inspection should be performed by a professional fibre
inspection organisation which is recognised by the National Fibre Inspection Bureau. The
organisation should produce a notary public inspection certificate, which is used as a wool
quality document for wool pricing and refund.

6. A wool supplier should pack the different types of wool into separate bales according to the
stipulation in the National Wool Standard and specify the production area, wool category and
grade, batch number and bale number. The supplier should send an application for notary
public inspection to a local professional fibre inspection organisation.

7. A wool buyer should request the notary public inspection certificate from the seller or
specify the necessity of an application for such inspection by a local professional fibre
inspection organisation in a trading contract.

8. In the case of disagreement on the inspection results from either of the two parties, the
seller and the buyer, an application for re-inspection should be sent to the wool inspection
organisation within 15 days after receiving the inspection certificate. Re-inspection on the
reserved specimens should be performed within 15 days after receiving the application. If the
disagreement is not resolved by the second inspection, a second application for re-inspection
should be lodged in a higher level of a professional wool inspection organisation within 15
days after receiving the re-inspection certificate. An application for the third inspection should
be sent to the National Fibre Inspection Bureau if the wool trading is conducted between
provinces and to the provincial fibre inspection organisation if the trading is within the
province.

9. When trading is on a small scale, professional fibre inspection organisations are responsible for supervision and inspection of wool quality and for providing technical guidance on how to comply with the stipulation in the National Wool Standard.

10. The organisation or person who applied for wool inspection should provide necessary assistance to the professional fibre inspection organisation and provide the required fee for notary public inspection service.

11. Professional fibre inspection organisations which are authorised by the National Supervisory Authorities of Technologies are responsible for dealing with the following illegal behaviour if they occur during wool trading:

(1) In the event of wool selling without application for notary public inspection and cheating in wool category and grade, the accounts should be re-settled based on the inspection results obtained. The seller should refund any overcharge. A fine of less than 10% of the total price of the trading wool may be applied. In severe cases or where the seller has been engaged in previous illegal behaviour, a 10 to 20% fine can be applied.

(2) In the event of wool pricing not based on the net wool weight during a large-scale trading, the responsible party may be fined up to 10% of the total price of the trading wool.

(3) In the cases where inferior-quality wool or non-wool materials have been added into the traded wool and production of false notary public inspection certificates or other illegal behaviour, the seller should provide any economic loss as a result of his illegal behaviour, return any illegal profit, and be fined for 10 to 20% of the total price of the traded wool. If a loss of wool industrial value results from the illegal behaviour, the seller should provide any economic loss, return any illegal profit and be fined 20–50% of the total price of the traded wool. In severe cases where a crime has been committed, a legal punishment against the crime is to be applied.

12. Professional fibre inspection organisations responsible for inspection of the primary processing product of wool execute the penalty for any illegal behaviour on the basis of inspecting results according to the stipulation for required product quality.

13. In the case of disagreement over the penalty to be applied, an application for re-judgement of the case can be made to the professional fibre inspection organisation which made the decision within 15 days after receiving penalty notice and can also sue in the court.

14. A disciplinary sanction is applied to anyone in professional fibre inspection organisations who engages in the following illegal behaviour:

(1) An economic loss as a result of his/her improper duty.

(2) Producing false results or changing inspection results.

(3) Malpractice for the purpose of his/her own benefit, improper use of professional powers and asking for or accepting bribes.

In severe cases where a crime has been committed, a legal punishment against the crime is to be applied.

15. The National Supervisory Bureau of Technologies is responsible for interpretation and public awareness of this document.

16. This document was implemented from 1 June 1993.

Appendix H

Quality Standard of Auctioned Wool

(An edition for trial implementation drawn up in March 1991 by the
Ministry of Textile Industry and the Ministry of Agriculture)

Introduction

Since the first experimental wool auction held in Urumqi in 1987, and along with the continuous expansion of the wool market, wool auctions have aroused great attention and the support of more and more people in the industry and animal husbandry sectors. In 1989 when the 40th anniversary of the founding of the People's Republic of China was celebrated, the "First National Exhibition of Sheep and Goat Breeds, Wool and Cashmere and Their Products" was organised. We summarised the previous work, reviewed some of the written implementation methods, organised experts of related departments and compiled the "Work Regulations for Wool Auction" which is intended for reference and use by people involved in the wool auctions. It is hoped that constructive suggestions will be initiated for future amendment and even more perfect and rational work regulations will emerge.

In April 1990, the Working Group of Wool Auction Reformation Pilots was formally established and the "Work Regulations for Wool Auction" was further amended. A revision of the "Work Regulations for Wool Auction" was passed via discussion at the "Conference of the Coordination Group for Wool Production Bases" held in May 1990 in Wuhan, Hubei Province.

On the basis of the suggestions made from various parts of China, the Working Group of Wool Auction Reformation Pilots invited representatives of both wool-supplying and wool-purchasing institutions for the revision of the regulations in Huhehot, Inner Mongolia in March 1991. Through thorough discussions, the regulations were revised, supplemented and renamed: "Quality Standard of Auctioned Wool".

During the process of the drafting of the standards, the current domestic production status was taken into consideration. Some time is needed before the conditions are right for a perfectly-run auction system to emerge in China. Therefore, on the basis of the principle of "seeking common grounds on major issues while reserving differences on minor ones", the standards should not only protect the initiative of animal production while leaving a wide margin to let people strive to achieve the objective via efforts in a short period of time, but also should be made more acceptable to industry and initiate a more attractive force in the market.

The institutions which took part in the drafting work included: Production Commission of the State Council, Ministry of Textile Industry, Ministry of Agriculture, National Bureau of Fibre Inspection of China, Shanghai Academy of Textile Sciences, Xinjiang Wool Research Institute, Xinjiang Textile Industry Supply and Marketing Company, Inner Mongolia Textile Industry Company, Inner Mongolia Animal Improvement Station, Nanjing Branch of the China Textile Goods and Material Company, Department of Raw Materials of the Jiaxing Textile Industry Company, and Shanghai Wool Top Company.

General principles for implementation

(I)   Before wool can be sold at the auction market, it should undergo the following process: grading and sorting, objective measurement and pricing on a clean-wool basis.

(II)   In objective measurement of wool and pricing on a clean-wool basis (appraisal of conditioned weight), legal notary appraisal should be implemented.

1. The supplying institution (hereinafter referred to as the supplier) which would like to take part in the wool auction should submit an inspection application to the local professional fibre inspection institution. One inspection is effective.

2. The local fibre inspection institution should use the method of objective measurement stipulated in the Quality Standard of Auctioned Wool, issue an appraisal certificate, scale the wool bales one by one and provide samples for display at the auction market.

3. The fibre inspection institution is responsible for the supervision and instruction during the whole process of the wool grading and sorting required for inspection.

(III) After a bargain is struck, loan for the settlement of accounts should be carried out on the basis of clean-wool price and the conditioned weight of clean wool.

(IV) Before the actual auction takes place, the presiding institution of the auction should announce the site, date of the auction market, and the types, specification and amount of wool according to the source of wool. It should also display the wool samples of labelled types and specifications. Auction can be conducted in accordance with the displayed samples, appraisal certificate and scaled conditioned weight.

(V) The buying institutions (hereinafter referred to as the buyer) should be enterprises (companies) who have the certificate in wool raw material marketing, and production units producing woollen textile products.

(VI) The personnel sent by the buyers should be familiar and knowledgeable with wool and have the ability to make selection decisions by themselves. They should submit a registration application accompanied with the entrust letter of the legal person to the auction market and obtain the "Qualification Certificate of Wool Buyer".

(VII) The wool purchased at the auction market should be used directly in production or be distributed to production units within the system. It is forbidden for the wool to be sold for profit. If such sale is found, the Qualification Certificate of Wool Buyer is revoked accompanied by criticism via bulletins.

(VIII) The "Technical Supervision and Arbitration Group" should be established in order to supervise the auction deals and arbitrate on disputes.

(IX) The research project of the "Modernisation of Wool Technical Management" should conduct investigations so as to further supplement the Standards. The grading and sorting, objective measurement and appraisal of conditioned weight for pricing wool on a clean-wool basis at wool production bases will continue to be taken care of by the institutions involved in this research project.

(X) The right of explanation of the Quality Standard of Auctioned Wool belongs to the "Working Group of Wool Auction Reformation Pilots".

A. QUALITY STANDARDS

These standards are suitable for domestically-produced fine wool, semi-fine wool and improved wool which has been graded and sorted for sale at the auctions.

1.   Technical specifications (Tables H1(a), (b) and (c))

Table H1(a) Technical Specifications for Fine Wool

Quality measures	Fine wool count			
	70	66	64	60
Fineness				
Mean fineness (μm)	18.1–20.0	20.1–21.5	21.6–23.0	23.1–25
Rate of coarse-cavity wool (%)	0.0	0.0	0.0	0.0
Discreteness coefficient	24	25	27	29
Staple length				
A: Mean	More than 8cm			
A: Minimum	6cm			
B: Mean	6–7.9cm			
B: Minimum	5cm			
C: Mean	4–5.9cm			
C: Minimum	4cm			
D: Mean	Less than 4cm			
D: Minimum				
Quality grade				
I	Refer to Section 1.4.1.1			
II	Refer to Section 1.4.1.2			
III	Refer to Section 1.4.1.3			

Note: When the grade and sort cannot be described by the technical specifications, actual samples should be used to judge the grade and sort.

Table H1(b) Technical Specifications for Semi-Fine Wool

Quality measures	Semi-fine wool count					
	58	56	50	48	46	44
Fineness						
Mean fineness (μm)	25.1–27.0	27.1–29.5	29.6–32.5	32.6–35.5	35.6–38.5	38.6–42.0
Rate of coarse-cavity wool (%)	0.5	0.5	0.75	0.75	0.75	0.75
Discreteness coefficient	31	31	32	32	32	32
Staple length						
A: Mean	More than 10cm					
A: Minimum	7cm					
B: Mean	7–9.9cm					
B: Minimum	6cm					
C: Mean	4–6.9cm					
C: Minimum	4cm					
D: Mean						
D: Minimum						
Quality grade						
I	Refer to Section 1.4.2					
II	Refer to Section 1.4.2					
III	Refer to Section 1.4.2					

Note: When the grade and sort cannot be described by the technical specifications, actual samples should be used to judge the grade and sort.

Table H1(c) Technical Specifications for Improved Wool

Quality measures	Improved wool		
	Sort I	Sort II	Sort III
Fineness			
Mean fineness			
Rate of coarse-cavity wool (%)	1.0	3.5–4.0	6.0–7.0
Discreteness Coefficient			
Staple length			
A: Mean			
A: Minimum			
B: Mean			
B: Minimum			
C: Mean			
C: Minimum			
D: Mean			
D: Minimum			
Quality grade			
I		Refer to Section 1.4.3.1	
II		Refer to Section 1.4.3.2	
III		Refer to Section 1.4.3.3	

Note: When the grade and sort cannot be described by the technical specifications, actual samples should be used to judge the grade and sort.

1.1 Classification of wool types

    1.1.1 All homogeneous wool could be classified into fine wool and semi-fine wool with 10 sorts of quality count on the basis of quality number.

    1.1.2 Improved wool is classified into three sorts on the basis of rate of coarse-cavity wool and quality characters: Sort I, Sort II and Sort III.

1.2 Wool fineness and rate of coarse-cavity wool

    1.2.1 Mean fineness and rate of coarse-cavity wool are the grounds for classification.

    1.2.2 Fineness discrete coefficient is used for reference in classification.

    1.2.3 Fine wool and semi-fine wool are not allowed to contain withered wool, kemp or coloured wool.

    1.2.4 Fleece after skirting should be measured according to the quality count of the main fleece. It is allowed to contain some wool of one higher grade and one lower grade, but the mean fineness and rate of coarse-cavity wool must accord to the corresponding quality number of the grade being assigned.

    1.2.5 Body wool piece and loose wool must be arranged to meet the requirement of the quality count of this sort and the quality count should be evaluated on the basis of mean fineness and rate of coarse-cavity wool.

    1.2.6 The rate of coarse-cavity wool is used as the basis for classification of improved wool.

    1.2.7 After grading and sorting, improved wool is classified by combination of the rate of coarse-cavity wool and quality. It is allowed to contain some wool of one sort higher or one sort lower wool.

1.3 Staple length

    1.3.1 Mean length refers to the natural length of the staple.

    1.3.2 Staple shorter than the specified minimum length should be picked away.

1.4   Quality grade
Wool is classified into three grades on the basis of quality characters and apparent morphology: Grade I, Grade II, Grade III.
 1.4.1   Fine wool
  1.4.1.1 Quality Grade I
  All the wool should be natural white homogeneous wool with fine structured staple, dense fleece, good and uniform fineness and length, and medium grease and suint. The colour of grease and suint should be white or milky white; hand feel should be soft with good elasticity with normal crimpness, uniform, clear, and a clean-wool rate of more than 45%.
  1.4.1.2 Quality Grade II
  All the wool should be natural white homogeneous wool with fine structured staple, good and uniform fineness and length, and medium grease and suint. White or milky white should be dominant in the colour of grease and suint; hand feel should be soft with good elasticity, normal crimpness and a clean-wool rate of about 40%.
  1.4.1.3 Quality Grade III
  All the wool should be natural white homogeneous wool. The structure of the staple is not very dense with slightly poorer uniformity in wool fineness and length, soft hand feel, slightly poorer elasticity, slightly abnormal crimpness and with wool tips on fleece top.
 1.4.2   Semi-fine wool (no grade for the time being)
  All the wool should be natural white homogeneous wool, with fairly good uniformity in wool fineness and length, large and shallow crimpness, and good lustre and elasticity.
 1.4.3   Improved wool
  1.4.3.1 Quality Grade I
  All the wool should be natural white, fundamentally homogeneous wool, consisting of fine down hair and coarse down hair with loose staple. The uniformity of fineness and length as well as crimpness, grease and suint and elasticity are very close to fine wool. However, the wool contains trace amounts of coarse wool and cavity wool but is without withered wool and kemp.
  1.4.3.2 Quality Grade II
  All the wool should be natural white heterogeneous wool. Fine down hair, coarse down hair and heterotypic fleece are dominant with small quantities of coarse wool and cavity wool, as well as trace quantities of withered wool and kemp. The crimpness is not obvious with small hair plaits or fine and long hair plaits.
  1.4.3.3 Quality Grade III
  All the wool should be natural white heterogeneous wool, consisting mainly of fine down hair, coarse down hair and heterotypic fleece. The wool contains fairly large quantities of coarse wool and cavity wool, small quantities of withered wool and kemp. The wool has the character of improved wool but is not free of the morphology of native wool, with coarse and long hair plaits.
1.5   Wool which has been graded and sorted should not mixed with the following wool
 1.5.1   Various defects of wool: cotted wool, skin-piece wool, dirty-clot wool, pitch wool, paint wool, scabby wool, tender wool, yellow-stained wool, and other edge-trimmed wool.
 1.5.2   Wool from the head, leg and tail of a sheep.
 1.5.3   Coloured wool.
 1.5.4   Raw-skin-clipped wool, grey faded wool.

B. PRE-SALE CLIP PREPARATION

1. Shearing

1.1   Shearing should be carried out on the shearing table. Earth, grass or other sites which will influence wool quality should not be used as the shearing site.

1.2   The sheep farm (or household of sheep husbandry) should divide the flock into groups for shearing in accordance with the feeding conditions and breed. This practice is beneficial for grading after shearing and can guarantee the uniformity of wool quality.

1.3   Shearing should be conducted according to operation schedules. The wool of the head, leg and tail of a sheep and contaminated wool should be clipped away first and stored separately. When shearing, it is essential to maintain the completeness of the fleece and care should be taken to avoid double-clip wool, and the remaining stubbles should be as short as possible.

2. Fleece edge sort-out

2.1   The whole fleece is spread with wool tips upward. An overall observation should be conducted so as to sort out fleece edges.

2.2   Wool contaminated with excretions, short wool, contaminated wool, and coarse-cavity wool at the edges of the fleece should be cleared away.

2.3   Fleece edge sort-out should be appropriate. After edge sort-out, cotted wool, pitch wool (or paint wool), concentrated seedy wool, thorny wool, skin-piece wool, coarse-cavity wool, coloured wool and other deteriorated wool on the fleece should be picked up and stored separately.

2.4   In cases of large areas of cotted wool, urine-yellow wool, tender wool, or wool with large quantities of thorns and coloured wool found in the fleece, the whole fleece should be put away separately for individual treatment.

2.5   The wool from the head, legs and tail of a sheep which is sorted out should be stored according to its length.

3. Grading and sorting

3.1   Staple samples should be randomly taken from the side, shoulder and rump of the main fleece after edge-sorting. The fineness and quality grade characters should be observed visually and the natural length measured so as to determinate the grade and sort.

3.2   The fleece fineness is determined according to the dominant quality number of the main body of the fleece.

     3.2.1   For example, if the quality number of the main body of a fine wool fleece is 64, it is allowed to contain wool of the fineness of one scale higher or one scale lower, such as 66 and 60. However, it is not allowed to contain wool of the fineness of more than one scale higher or one scale lower, such as 70 and 58. The rule for fleeces of other counts can be determined by analogy and the same theory is applicable to semi-fine wool.

     3.2.2   As for improved wool (referring to improved fine wool and improved semi-fine wool), it is allowed to contain wool of an adjacent grade to the main body wool grade, but cannot contain wool beyond one grade difference.

3.3   The sort of fleece length is classified according to the mean of fleece natural length.

All wool shorter than the minimum length of the technical standard specifications should be picked away.

C. APPRAISAL OF CONDITIONED WEIGHT

Wool which is to be sold at auction should be shipped to a storehouse in batches according to Section E of this standard. The appraisal institution will scale the wool bales by checking them against the submitted form and take appraisal samples at a pre-defined rate.

1. Technical specification

1.1 The maximum weighing weight should be 500kg with the minimum graduated reading of 0.5kg.

1.2 Before weighing, the scale should be examined visually. After moving the scale for a certain distance, the scale should be placed on a horizontal level site. The four legs must be on solid ground, with accessory parts in normal contact. Scales without wheels should not be moved.

1.3 During the process of appraisal, the scale should not be moved at will and should be maintained at a persistent balance.

1.4 The precision of scales used in wool measuring should be checked by the local measurement department. It is required to recheck the scales before the use of repaired scales. It is essential to follow strictly the unified operation rules.

2. Data processing

2.1 The scaling forms of the sequential bale number of the whole batch of wool should be worked out on the basis of the scaling records, and serve as the conditioned weight of clean wool.

2.2 The recorded figure should contain one decimal. In the case of one digit approximation, please refer to "GB 8170 Data Value Approximation Rule".

D. OBJECTIVE MEASUREMENT

1. Sampling method

1.1 Sampling includes two aspects; a drilled sample for objective appraisal and a display sample for auction.

1.2 A drilled sample should have a total weight of more than 750g for the measurement of scouring yield, fineness and vegetable matter content.

1.3 As for the display sample, wool bales should be opened at both ends or randomly take enough sample from the middle of the bale. When the batch is less than 10 bales, samples should be taken from each bale. When the batch is less than 30 bales, samples should be taken from 10 bales. In cases of a batch of more than 30 bales, each 20-bale increment should add one more sampled bale. As a whole, the total sample weight of a batch should be more than 6kg, specifically for wool length appraisal and auction display.

1.4 Drill sampling method

    1.4.1 Drill sampling should be conducted simultaneously with weighing, so as to guarantee an unchanged conditioned weight of wool bales.

    1.4.2 The number of bales needed for drill sampling per batch is listed in Table H2. Each appraisal batch should drill at least 32 holes.

Table H2 Number of Bales Needed for Drill Sampling per Batch (Examples Only)

Number of bales per batch	Less than 32 bales	50	75	100	150	200	300	400	500
Number of bales needed for drill sampling	Whole batch	48	56	60	69	73	78	80	82

1.4.3   The sample drilling apparatus has a diameter of 25mm, and is pressed into wool bales either manually or mechanically.

1.4.4   Before drilling, the packing material of the wool bale should be cut open so as to prevent the packing material from entering the drilled sample. The drill sampler should enter the wool bale vertically against the pressing direction.

1.4.5   The drill-sampler's insertion site must be a random position at the top or bottom of bale. The distance from the sampling site to the bale edge should not be less than 100mm. The drill holes must alternate between the top and bottom of the bales.

1.4.6   The depth of drill holes should be more than 90% of the drilling tube.

1.4.7   The weight of drilled samples should exceed the weight of five sub-samples, with 150g in each sub-sample and a total weight of 750g.

1.4.8   When the batch is less than 32 bales, each bale should be drill-sampled by one or two holes, so as to ensure a total of 32 holes in the batch. When a batch has too few bales and one or two holes in each bale could not reach the total sample weight of 750g, it is suggested that more holes be drilled in each bale so as to meet the sample weight requirements. However the distance between holes should be at least 50mm.

1.4.9   Samples should be put into plastic bags and sealed immediately after drill-sampling, so as to keep the wool, soil and foreign material intact, to prevent samples from being exposed to air, and to guarantee the original moisture content of the wool.

1.4.10 Wool samples should be weighed immediately after drill-sampling, with batch number, weight, wool name, date and place of sampling, sampling person, etc., recorded. If a batch of wool bales are drill-sampled separately in the morning and in the afternoon, or over two consecutive days, the weighing should be done separately and the weight should be added up.

2.   Scouring yield

2.1   The wool sent for appraisal must be from the same producing area and must have been edge-trimmed, graded and sorted and have the same packing. There should be no breakage in the bales and the wool bales should be of the same size, approximately the same weight, have an even distribution of clean-wool percentage, and be randomly packed.

2.2   Wool expansion

2.2.1   All the wool sampled and weighed should be expanded via a medium wool expansion machine in one process.

2.2.2   The feeding of the wool into the wool expansion machine should be even and maintained at a speed of 50g per minute. Care should be taken to avoid feeding too much at a time so as to avoid blocking the wool transmission tube. In case of big blocks of excretions, they should be picked up and cannot be fed into the machine.

2.2.3   The wool box should be cleaned every 5 minutes by putting the expanded wool into a plastic bag or air-tight container, and then continue the feeding operation.

2.2.4   After the completion of one batch of wool samples, open the cover of the machine, check the beater, nail, axle, screen bottom, etc. and pick up the wool remaining in the machine residues and put them in the expanded wool sample of the same batch. Care should be taken not to keep any residual wool in the wool expansion machine before running the machine for the next operation of another batch of wool sample.

2.2.5   Put the expanded and mixed wool sample on the appraisal table, take sub-samples (about 120–130g in each sub-sample) by taking wool from different sites. A total of five sub-samples should be taken, of which three sub-samples are used for clean-wool percentage appraisal, two sub-samples are used as preparatory samples and the rest of the sub-samples should be weighed and recorded separately.

2.2.6   The foreign material left during wool sample expansion must be collected and weighed. The actual tested wool sample weight of each sub-sample is calculated according to the following formula:

W = Sub-sample weight taken after expansion (about 130 g)

$$X = \frac{\text{Total weight of drilling sample}}{\text{Weight of 5 sub-samples + Weight of remaining sample}}$$

The calculation should be accurate to 2 decimal places.

2.2.7   The raw wool sample should be weighed by using a balance sensitive to 0.1g. During the process of wool expansion, loss of weight is not allowed with the exception of losing soil and foreign materials.

2.3   Wool scouring

2.3.1   Wool-scouring equipment: Wool scouring should use a pressure wool-scouring machine equipped with a slot board and water drainage holes. There should be two wool-scouring cases and a copper screen (60 meshes) filter and appropriate drainage system.

2.3.2   Wool-scouring technology: In the first trough, the detergent   should be able to clean one wool sample. The second and third troughs should be able to rinse three wool samples (Table H3).

2.3.3   Detergent formulation: LS detergent should first be formulated into 1% concentration solution for scouring wool. Due to the fact that LS detergent is not readily soluble and is easy to clot, it is suggested that the LS detergent powder should not be poured directly for use.

2.3.4   Operation procedure for wool scouring:

2.3.4.1 Pull the handle located at the centre of the cover of the wool-scouring machine, take down the seal cover, drain away the remaining water, turn tight the screw of the drainage orifice and pour in warm water to the specified volume. Add in LS detergent, Yuanming Powder, etc.

2.3.4.2 Measure the temperature of the scouring water, put in the appraised wool sample when the temperature of the liquid reaches 57–60°C, put on the air-tight cover when the wool sample has been thoroughly soaked,   turn tight the screw of the cover, then pull down the handle located at the centre of the cover   so as to tighten the cover.

2.3.4.3 Soak for 2 minutes. (You could also make use of the remaining detergent of previous scouring to soak the wool sample in another trough or basin. This will achieve an even better cleaning result.)

2.3.4.4 Rotate the handle clockwise at the speed of 1 rotation per 2 seconds for 1 minute and then rotate the handle counter-clockwise for 1 minute. Under conditions of micro-pressure detergent and mechanical rotating friction, the lanolin of the wool and soil in the wool as well as other foreign materials in the wool are removed.

2.3.4.5 In case of very greasy wool or particularly difficult-to-clean wool, it is suggested to increase the rotating speed to 50 rotations per minute and to rotate 1 minute both clockwise and counter-clockwise.

2.3.4.6 After completion of cleaning in the first trough, pull the handle upright, turn the cover counter-clockwise for 90 degrees, release the pressure, turn open the cover of the water cover and release the detergent from the machine. Use the copper screen (60 meshes) filter for filtration. (The released detergent could be used to substitute the soaking solution of the first trough.)

2.3.4.7 Turn tight the cover of the water drainage, add in 12–15 litres of clean water of 40–45°C, turn tight the upper cover, rotate the handle at the speed of 30 rotations per minute clockwise for 1 minute and counter-clockwise for another minute. This is the procedure for the second trough. A similar procedure should be repeated for the third trough.

2.3.4.8 Another method is to take the wool sample from the first trough (the wool is not allowed to be retained in the container) and rinse the wool sample by following the procedures listed in Table H4 for the second trough and the third trough (or basin).

Table H3 Wool-Scouring Technology

Technology	First trough	Second trough	Third trough
Wool-scouring equipment	Pressure wool-scouring machine	Scouring case and trough (basin)	Scouring case and trough (basin)
Detergents:			
LS detergent	0.1%	Clean water	Clean water
Yuanming powder	0.5%	Clean water	Clean water
Temperature (°C)	55–60	40–45	40–45
Duration (minutes)			
Soaking	2	2–3	2–3
Shaking	2	2–3	2–3
Volume of water (litres)	12–15	12–15	12–15

Table H4 Wool-Cleaning Technology

Technology	Trough 1	Trough 2	Trough 3	Trough 4
Wool-scouring equipment	Scouring case and trough (basin)			
Detergents:				
LS detergent	0.08%	0.05%	Clean water	Clean water
Yuanming powder	0.5%	0.3%	Clean water	Clean water
Temperature (°C)	55–60	55–60	40–45	40–45
Duration (minutes)	3	3	3	3
Volume of water (litres)	15	15	15	15

2.3.4.9 When using a pressure wool-scouring machine, the technology of the four-trough wool cleaning procedure listed in Table H4 for wool cleaning should not be used. In the first trough and second trough, the detergent could be used to clean two wool samples while in the third and fourth trough, the clean water could be used to rinse three wool samples.

2.4 Drying

2.4.1 Put the cleaned wool sample into a nylon bag of over 100 meshes for water drainage so as to prevent loss of wool scraps.

2.4.2 After the water drainage, the wool sample should undergo pre-oven drying or air drying.

2.4.3 Tear loose the pre-oven-dried wool sample or air-dried wool sample.

2.4.4 Put the wool sample in an oven of 105±2°C for oven drying. Weigh and record the weight of the wool sample 3 hours after the oven drying. (The scale for weighing in the oven should have a sensitivity of 0.01g.) When the temperature returns to normal, weigh the wool sample every 20 minutes until the difference between the two weights does not exceed 0.05% of the final weighing. The final weighing weight is used as the oven-drying weight (D).

2.5 Scouring yield calculation:

2.5.1 Formula 1 (suitable for wool samples which have less than 2% of vegetable matter content):

$$Y_1 = \frac{D \cdot (1-R)}{G} \cdot 100 \ (\%)$$

2.5.2   Formula 2 (suitable for wool samples which have more than 2% of vegetable matter content):

$$Y_2 = \left[\frac{D \cdot (1-R)}{G} - \frac{V - 2}{100} \cdot 1.5 \right] \cdot 100 \ (\%)$$

where Y = scouring yield of examined wool sample (%)
$\qquad$ D = oven-drying weight of examined wool sample (g)
$\qquad$ R = rate of official moisture regain (%)
$\qquad$ V = content of vegetable matter (%)
$\qquad$ G = actual weight of raw wool under appraisal (g).

Note: The rate of public moisture regain of fine wool is temporarily assumed to be 16% while that of semi-fine wool, improved wool and native wool is 15%.

2.5.3   Rounding of calculation: Calculate the scouring yield of three wool samples and take the mean of them as the scouring yield of the whole wool batch. The calculation should be rounded to 2 decimal places.

3.   Fineness (drilling wool sample, projection method)

3.1   Measuring equipment and material

Microprojector, objective-lens micromeasurement ruler, glass slide for carrying material, glass slide for covering, glycerol (or paraffin), dish, fine glass stick, single-side blade (or H's slice equipment), wedge ruler and tweezers.

3.2   Test method

3.2.1   Take representative small samples from the cleaned and oven-dried wool sample. After a balance period under standard temperature-humidity conditions, use two single-side blades to cut fibres of 0.1–0.2mm, put them in the dish, drop in the appropriate amount of glycerol, and mix the liquid thoroughly until the single fibres suspend on the surface of the liquid. Then use a fine glass stick to move the liquid on to the glass slide for carrying material and cover it carefully with the glass slide for covering. Three such glass slides should be prepared for each wool sample.

3.2.2   Use the wedge ruler and objective-lens micromeasurement ruler to adjust the magnification fold of the projector to 500 folds.

3.2.3   Put the glass slide on the carrying deck so as to let the fibre be projected on the white screen on the surface of a table, adjust the focus until the picture of the fibre is clear and measure the fibre fineness via the wedge ruler.

3.2.4   Do not skip but repeat the measurement of fineness of the fibres. The measurement should start from one corner of the glass slide, from left to right and from top to bottom. In the case of a fibre which has great variation in fineness, measurement should be conducted in the middle of the fibre. The unclear and unobvious fibres or overlapping fibres should not be measured. It is recommended to measure the fibres in the projection centre where the magnifying fold is correct. Make one side of the wedge ruler overlap with the fine line of the outline of one side of the fibre, let another side of the wedge ruler match with another side of the outline of the fibre, and make a mark so as to measure the fineness of a fibre. Repeat such measurements until the specified number of fibres is met.

3.2.5   The specified number of fibres that should be measured for each slide are:
Fine wool: 300 fibres.
Semi-fine wool and improved wool: 400 fibres.

3.2.6   The specification for the difference of two slides: The difference of two slides of domestic quality count wool (fine wool, semi-fine wool), and improved wool should not exceed 3% of the means of the two slides. The difference of two slides of native wool should not exceed 4% of the means of the two slides. Otherwise, another slide should be measured and the mean of the three slides should be used as the final fineness of the wool.

3.2.7   Mean of fineness ($\bar{x}$), standard deviation (σ) and coefficient of variation (CV) are calculated via the following statistical method:

$$\bar{x} = A + C \cdot T \qquad (\mu m)$$

$$\sigma = \sqrt{\frac{\Sigma d^2}{N} - C^2 \cdot T^2} \qquad (\mu m)$$

$$CV = \frac{\sigma}{\bar{x}} \cdot 100 \qquad (\%)$$

where A = assumed mean
        T = number of classes

$$C = (\frac{f_i d}{\Sigma f}) \quad \text{= weighted distance between class means}$$

   f_i = number of fibres in each class i
   N = total number of fibres measured
   d = distance between class means measured in microns.

Note: If the mean fibre diameter exceeds 52.5μm the wool is regarded as coarse wool. Coarse wool with medullated fibres is regarded as coarse medullated wool.

4.   Staple length

4.1   Measuring equipment: steel measuring tape of 150mm length and black flannel board.
4.2   Operational procedure: Randomly take 300 complete staples from the whole batch of wool, and put the staples flat on the black flannel board. Keep the staple in a natural state of crimpness. It is not allowed to stretch the staple or change the natural crimpness of the staples. The staple length should be measured on the tip of the staple (excluding the palpus) and the measured results should be carefully recorded.
4.3   The mean natural staple length is calculated according to the following formula:

$$\bar{x} = \frac{\Sigma x}{300} \qquad (mm)$$

where $\bar{x}$ = mean of staple natural length (mm)
        x = the measured natural length of each staple (mm).

5.   Vegetable matter content

5.1   Measuring equipment and material, copper screen (60 meshes) filter, oven dryer, other glassware, stainless steel beaker, plastic bottles, sodium hydroxide, and phenolphthalein test reagent.

5.2   Operational procedure
       5.2.1   Take 40g of wool from the wool sample, which has been examined for scouring yield, to measure the vegetable matter content. Take another two lots (5g each) for the

moisture test. The oven drying is the same as that described in paragraph 2.4. Then calculate the moisture content.

5.2.2   Pour 5% sodium hydroxide into a stainless steel beaker at the bath ratio of 1:15 (about 600ml), put the beaker on an electric stove or gas stove to boil, put 40g of wool sample into the boiling sodium hydroxide solution, maintain the temperature and stir for 3 minutes. Add cooled distilled water to dilute the solution of sodium hydroxide after the wool has been completely dissolved.

5.2.3   Use a copper screen of 60 meshes to filter the diluted solution, rinse it continuously with clean water until the sodium hydroxide is completely washed away. Continue the rinsing operation by using distilled water in bottles until the rinsing water does not turn red when applying the phenolphthalein test reagent.

5.2.4   Pick paper scraps, rope, sheep-skin bits, plastic crumbs, etc. out of the sample.

5.2.5   Put the remaining material into a bacterial funnel, move the funnel into an oven of 105±2°C for drying until the weight of the sample is constant.

5.2.6   The content of vegetable matter is calculated according to the following formula:

$$V = \frac{\text{Dry weight of vegetable matter}}{\text{Weight of tested sample (about 40g)} \cdot (1-W)} \cdot K \cdot 100(\%)$$

where: V = content of vegetable matter (%)
W = moisture content of wool sample (%)
K = corrected co-efficient = 1.1.

6.   Certification of objective measurement

The report form should contain the following items:
(1)   Batch name, quality count or grade and sort, and producing area of the wool.
(2)   The total number of bales of the whole batch and the weight of raw wool (gross weight and net weight).
(3)   The final results of measured scouring yield and the conditioned weight of clean wool of the whole batch of wool.
(4)   The number of drilled bales in the batch as well as the number of opened bales for wool sampling.
(5)   Mean fineness of wool (in μm) and co-efficient of variance (%).
(6)   Mean staple length (mm).
(7)   Content of vegetable matter (%).

E. PACKING, MARKING, STORE AND TRANSPORTATION

1.   Packing

1.1   Wool which has been skirted and classified should be baled separately according to the destination and grade. Bales should be made of flax cloth or white cotton cloth. Bales of wool should be pressed bales.

1.2   The wool bales should be of uniform sizes, with dimensions of 800cm by 400cm by variable height. The net weight of wool in each bale should be in the range of 80–100kg.

2.   Marking

2.1   The suppliers should clearly mark bales with an ornamental engraving template and in dark paint. The content of these marks should include: (1) name of the farm, (2) batch number, (3) name of wool and grade and sort, (4) registered number of the wool classer, (5) gross weight and (6) bale number. Wool bales without such marks, in general, are not allowed to be auctioned.

2.2   The arrangement of marks on the packed bales is as follows:

Name of farm	< - - - - - | - - - - - -	XX stud farm
No. in auction	< - - - - - | - - - - - -	911002
Grade and sort	< - - - - - | - - - - - -	64AII
		Woolclasser No: 47
		Gross weight: 102
		Bale number: 9

2.3   The auction number consists of 6 digits. The first two digits show the year in which the wool was produced. For example 91, 92, 93... The third digit shows the producing province or autonomous region. Please refer to the following table for details:

Code	Province or region
1	Xinjiang
2	Xinjiang Production and Construction Corps
3	Inner Mongolia
4	North east (Heilongjiang, Jilin)
5	Sichuan, Yunnan, Guizhou
6	East China
7	North west China
8	North China
9	Central China
0	Others

The fourth to the sixth three digits show the auction batch number, for instance: 001, 002, 003, etc.

A letter at the end of the batch number indicates that the wool belongs to one of the following categories:

Code	Wool type
G	Improved wool
X	Native wool
T	Tender wool
P	Skirted wool
C	Cotted wool
B	Seedy and thorny wool
H	Yellow-stained wool
L	Pitch and paint wool
K	Coloured wool
R	Head, leg and tail wool
D	Excretion-contaminated wool
O	Others

2.4   Wool bales of the same producing area (farm), same category and same grade or sort should be grouped together to form a batch. A batch size of 200–300 wool bales is appropriate.

3.   Store and transportation

3.1   Wool bales painted with marks of the same batch number and same wool category should be stored together by piling up in stacks. Irrespective of whether they are indoors or outdoors, the wool bales should be stored on rock, cement or wood pillars about 20–30cm above the ground. In cases of storing outdoors, the stack should be covered with tarpaulin, so as to protect the wool from getting wet and deteriorating.

3.2   During transportation, the wool bales should be loaded separately in accordance with the batch number, type and grade and sort. It is forbidden to load wool bales of different producing areas or different types in the same cargo wagon.

F. WORK REGULATIONS FOR AUCTION

1.   Qualification registration

1.1   Only people who have the certified qualifications or who possess the entrusted letter of a legal person can register and take part in auction activities.

2.   Form of auction

2.1   The auction market uses a method of public bidding. The deal will be final when there is no one to increase the price after the auction presiding person repeats a price three times and hits the final board with a wooden hammer.

2.2   When there is no one who wants to purchase the wool after the auction presiding person calls the start auction price, the auction market will negotiate with the supplier for adjustment of the start price and then commence the auction again. Another method is for the wool to be purchased at a protected price by the auction administration.

2.3   After the auction deal is final, the "Certificate of Deal" should be issued on the spot. The contract will be signed according to the "Certificate of Deal", and the display sample could be obtained which will serve as the sample for check-and-accept of the wool bales.

2.4   There are a total of five copies of one trade contract at the wool auction market. The contract will became effective as soon as it is signed by the legal representative of the recipient. One copy of the contract will be kept by the auction presiding institution, while the other four copies will be taken back to the institution of the recipient for stamping with the seal of the institution. It is required to send two copies of the contract to the supplier within 15 days.

2.5   The supplier and the recipient should negotiate whether the contract needs notarisation and verification. The responsibility of the cost of notary should be negotiated between the supplier and the recipient.

2.6   Both the supplier and the recipient of the auction should be faithful and adopt the contract.

2.7   The supplier should guarantee the uniformity between goods and sample and within a complete packing. The marks on the wool bales should be clear and the wool bales should be shipped to the recipient on time. It is not allowed to fake or use lower-grade wool to substitute for higher-grade wool.

2.8   The recipient should mail back the contract on time, pay the bargain money (deposit), organise a timely check-and-accept, and finish the payment for goods received within the time prescribed in the contract.

3.   Determination of prices

3.1   The wool auction price includes: wool purchase price, rate of difference between purchase and sale (including tax), grading and sorting technical service fee, baling fee (including material), inspection fee, transport fee from the stud farm to the collecting and distributing centres, quarantine and disinfection fee, transport loading and stacking fee, and market service fee, as well as loan interest, etc.

3.2   The quality differential ratio (for pricing) among different grades and sorts of wool: "64B2II" is regarded as 100% as against other grades and sorts of wool in the wool quality differential ratio (QDR) system for fine wool and improved wool.

3.2.1   Fine wool

3.2.1.1

Count	QDR
70	108%
66	104%
64	100%
60	96%

3.2.1.2   The staple length is further classified into B1, B2 and C1, C2 on the basis of the B and C sorts. The quality ratio of each sort is as follows:

	Mean staple length	QDR
A	≥ 8.0cm	109%
B1	7.0–7.9cm	104%
B2	6.0–6.9cm	100%
C1	5.0–5.9cm	90%
C2	4.0–4.9cm	75%
D	≤ 4.0cm	55%

3.2.1.3

Quality grade	QDR
I	104%
II	100%
III	97%

3.2.2   Improved wool

Quality grade	QDR
I	86%
II	62%
III	48%
Below III	33%

3.2.3   Semi-fine wool
Still under compilation.

3.2.4   Edge-trimmed wool

3.2.4.1 The edge-trimmed wool of fine wool is classified into 64 count, more than 64 count and 60 count in fineness, and B and C in staple length. The QDRs of each grade and sort are as follows:

Count	Staple length	
	Higher than B	Lower than C
64s and higher	75%	65%
60	70%	60%

3.2.4.2 The QDRs of edge-trimmed wool for improved wool:

Edge-trimmed wool higher than Grade II	53%
Edge-trimmed wool lower than Grade III	40%

3.2.5   The QDRs of other kinds of wool:

Head, leg and tail wool	40%
Autumn–summer wool	50%
Coloured spotted wool, raw-skin-clip wool	60%

The price of other defect wool is determined according to the actual quality.

3.3   Start price at auction: "64B2II" is regarded as the standard grade, with a QDR of 100%. The wool basic price (i.e. the price of "64B2II" wool) is reached on the basis of the regulations on the wool purchase price stipulated by the local government or the local price control department in the major wool-producing areas with reference to the wool-marketing situation. The start price at the auction of the wool grades is calculated according to their respective quality differential ratios.

3.4   Protect price (i.e. cost price): Wool which fails to make a deal at auction is purchased by the auction presiding institution at the protection price. The protection price is temporarily set at 90% of the start price at auction.

For the time being, there is no protection price for edge-trimmed wool, head–leg–tail wool and other defect wool.

3.5   In order to precisely and timely master the production and demand, and organise a timely marketing of wool, it is essential to organise an expert group to undertake a comprehensive appraisal of the wool quality against display samples and documents so as to fix a final start price before the auction.

4.   Methods of payment

4.1   Methods of settling accounts: The method to settle accounts should be negotiated between the supplier and the recipient on the basis of the stipulations formulated by the People's Bank of China.

4.2   After the auction deal, the party (either the supplier or the recipient) who violates the contract should take care of all the fees and expenses at the auction market, and is mandated to pay the fine which is about 10% of the total amount of the contract.

4.3   After the auction deal is made, the recipient should pay the bargain money which is about 30% of the contract. In cases of failing to pay the bargain money, the recipient should be responsible for the interest at a rate of 1.5% per month starting from the 16th day.

4.4   The recipient should start to ship the wool within 45 days upon receipt of the bargain money and sending of the receipt to the supplier. In cases of failing to ship the wool, the supplier should be responsible for the interest at a rate of 1.5% starting from the date of the receipt of the bargain money (with the exception of an alternative agreement between the two parties).

4.5   The recipient should make all the payment for goods within 10 days upon the reception of the shipped wool. In a case of failing to do so, the recipient should pay a fine at a rate of 0.03% on a daily basis in addition to the interest.

4.6   The auction market will charge a service fee at the rate of 2% of the total amount of the contract, and a production protection fund at the rate of 1% of the total amount of the contract which will be taken care of by the supplier and collected and paid on time by the supplier upon the receipt of the bargain money.

5.   Check-and-accept

5.1   When a supplier wants to sell wool at an auction market, they must first make an application to the local fibre inspection institution and acquire a certificate of wool inspection for auction (four copies for each contract) before they have the right to take part in auction activities.

5.2   The sorted wool of the same quality count or the same grade submitted for inspection should be weighed and drill-sampled as one batch and have one effective certificate for one auction batch at the auction.

5.3   The application for wool inspection should be submitted to the fibre inspection institution with documents of the wool bale batch number, category, type, number of wool bales, weight and scale form as well as other batch code-related documents.

5.4   The recipient should arrange check-and-accept and do the same as the supplier within the deadline for demanding claim indemnity.

5.5   There is a technical acceptable range for the results of check-and-accept between the supplier and the recipient.

Conditioned weight of clean wool	±1.5%
Mean fineness	±3.0%
Mean staple length	±0.5cm
Content of vegetable matter	0.2% (absolute value).

5.6   The check-and-accept of tender wool and other defect wool should be, in principle, done on the basis of the display samples. The sold wool of the following wool categories should not exceed the following percentage: tender wool 3%, other defect wool 5%.

5.7   The auction presiding institution is responsible for the preservation of the wool display samples. In a case of an imperfect match between the display wool and sold wool, the responsibility should be taken care of by the original inspection institution. In cases of repeated accidents like this, the qualification for wool inspection of the irresponsible fibre inspection institution will be suspended on the result of investigations thereafter.

6.   Arbitration of claims

6.1   The recipient should conduct the check-and-accept work on the basis of the methods of objective measurement stipulated by the Quality Standard of Auctioned Wool. In cases of a mismatch between the received wool and the displayed wool sample, or a mismatch of results between check-and-accept and the objective measurement certificate which exceeds the claim range, the recipient has the right to file a claim application within 45 days upon receiving the purchased wool. (This will be determined according to the latest postal mark.)

6.2   A claim report should clearly state the claim item and the check-and-accept result of the recipient, arrival date of the purchased wool and claim submission date. The recipient should inform the supplier, fibre inspection institution and arbitration group with one copy of the claim report for each of them. Upon the receipt of the claim report, the fibre inspection institution should organise a joint inspection in the presence of both the supplier and the recipient. The result of the joint inspection will be the basis for account settlement. If there is still a dispute, the recipient can file an application for arbitration.

6.3   Range of claims
      Conditioned weight of clean wool   =   ±2%
      Mean fineness   =   lower than the lower limit of the fineness
      Mean staple length   =   lower than the lower limit of the staple length
      Content of vegetable matter   =   more than 2%.

6.4   Claim stipulations
      6.4.1   When the conditioned weight of clean wool exceeds ±2%, the settlement of accounts should be calculated upon the results of the conditioned weight of the joint inspection. In the case of loss of raw wool weight, there is no separate arbitration transaction. There should be an integrated calculation upon the scouring yield.
      6.4.2   Mean fineness should be calculated upon the results of the joint inspection.
      6.4.3   Mean staple length should be calculated upon the results of the joint inspection. When the wool contains staples which are shorter than the minimum allowed for the grade, the proportion of this short wool is temporarily set at 3%. When the content exceeds 3%, a negotiation should be held between the supplier and the recipient on the basis of the actual status of the wool. Wool less than 4cm in length should be weeded out.
      6.4.4   In the case of wool with vegetable matter between 2 and 5%, the portion which exceeds 2% should be deducted from the conditioned weight of clean wool at the rate of 1:1.5. Wool which has more than 5% of vegetable matter is called seedy-and-thorny wool. No separate transaction should be done for the claim on vegetable matter.
      6.4.5   The check-and-accept of tender wool and other categories of defect wool is acted upon on the basis of the display samples. When the content of tender wool and other categories of defect wool exceeds that of the display samples, it should be solved through negotiations between both the supplier and the recipient. In case of arbitration, the sealed wool sample should be compared with the display wool sample held by the recipient.
      6.4.6   In the case of shortage of wool bales or rain-damage and other abnormal situations after receiving the purchased wool by contract, the recipient could file a claim to the insurance company according to the "safety and accident record" of the railway cargo service.

6.5   No claim is legally allowed in one of the following situations:
      6.5.1   Expiry of the effective period of claim.
      6.5.2   The wool bales have been opened and used which could not reflect the actual status of the original wool conditions. However, if the batch of wool has been mixed with large quantities of false grade wool, or mixed with short wool of less than 4cm in staple length, it will not be restricted by this regulation.
      6.5.3   The check-and-accept has not followed the appraisal procedures stipulated in the unified regulations.

6.6   Cost of joint re-inspection
      6.6.1   The cost arising from the joint inspection of the wool should be taken care of by the responsible party according to the regulations.
      6.6.2   The responsible party should take care of all the travelling expenses of the arbitration staff in addition to the payment of double the amount of the arbitration inspection fees.
      6.6.3   If the error in reality belongs to the inspection institution, after joint inspection, the inspection institution should take care of all the expenses arising on the joint inspection in addition to the responsibility and return of the original inspection fees.

7. Other precautions

7.1. In order to enter the auction market, it is required to wear badges of the auction market. The person who possesses the auction permit should attend in person at the auction; no representative should be entrusted to attend the auction bidding. If a representative is required, then notification should be made prior to the actual auction date so as to maintain the order and assure the safety of the auction activities.

7.2 When examining the wool samples, only visual examination is permitted. It is forbidden to touch or even turn over the displays at will, and it is absolutely forbidden to take away the displays so as to prevent mixture or loss of display samples or undermine the original quality status of the wool samples.

7.3 People who have a "Visiting Permit" are allowed to visit the displayed wool samples. They are only allowed to observe the auction bidding and have no right to respond to any price biddings.

7.4 The auction activities are of an open form. One minute after the declaration of the start price, bidding should be followed by holding up a plate. During the bidding process, a price will be repeated every 30 seconds. In case of no reply to a price after three repetitions, the auction presiding person will hit the board with a wooden hammer and the deal is made and the deal process follows on the spot.

7.5 At auction biddings, every wrong bidding or late bidding will be ineffective. After the hit of the wooden hammer, it is not allowed to hold up the plate for new auction prices.

7.6 At the auction hall, everyone should be quiet and keep the auction hall clean and tidy. It is not allowed to smoke, shout, gang together to press the price or raise the price at the auction. In the case of a need to offer opinions, hands must be raised before comments can be made.

G. GLOSSARY

1. *Auction market*: The site where the farmers entrust brokers to sell the wool via a middleman. Such auction markets, in general, are located in a port city, such as Sydney and Melbourne in Australia, Wellington and Napier in New Zealand, etc. The auction market will make public the auction agenda for the whole year. In general, there are two auctions per month during the busy season and once per month during the slack season. The auction date should not overlap adjacent auction market dates, so as to let more wool businessmen take part in the auctions and allow for competitive purchases. The market brokers will provide services for both the seller and the buyers. For instance, they will help the farmers to merge odd batches of wool for sale, scale the wool bales before entering the storehouse, store the wool bales, submit inspection applications, display wool samples, compile and publish auction catalogues, propagate wool marketing via advertisement, organise and preside over the auction activities, etc. They will also assist the buyers and sellers in contract signing, clearing payment for goods, retaining the tax, collecting administration and service fees (or service charge), and wool shipment transactions, and other things during the whole process of services.

2. *Trade market*: This refers to wool-marketing methods other than the auction, including the organisation of trade negotiation, regular market trading, the purchase of wool from the grass-roots unit in the wool-producing areas, the direct sale by contract between the farms and wool businessmen or factories, and other activities.

3. *Grading and sorting (pre-sale clip preparation)*: In general, this refers to operations which include the edge-trimming of contaminated wool, head-and-leg wool, belly wool, excessively short wool, seedy-and-thorny wool, etc. from the fleece to achieve uniform fleece quality, and then undertaking an integrated appraisal of the major part of the fleece on the basis of the

quality count (grade), staple length and grade. This should be carried out simultaneously with the shearing operation on stud farms.

4. *Objective measurement*: Objective measurement refers to the use of scientific instruments to examine wool and to describe wool quality according to scientific measurements. This is an antonym of the traditional optical estimation method of wool grade evaluation which is also called subjective measurement. On the basis of the technical conditions in China at the present time, the items of objective measurement include wool fineness, staple length, scouring yield, and content of vegetable matter. This is a central key link in the whole process of the implementation of the modernised technical administration in the construction of wool production bases.

5. *Appraisal of conditioned weight*: This is undertaken by the fibre inspection institutions on the basis of an application form for wool inspection. The fibre inspection institution is responsible for checking the painted marks and sequence number on the bales, for performing the weighing bale by bale, and producing the check certificate for the wool batches. This operation is one of a sequence of work procedures in drill-sampling for measuring scouring yield.

6. *Scouring yield*: This refers to the percentage that the clean-wool weight represents of the weight of raw wool after scouring the raw wool and allowing for the official moisture regain.

7. *Conditioned weight of clean wool*: This refers to the weight of clean wool after allowing for the official rate of moisture regain after scouring. During check-and-accept, the product of the raw wool weight (excluding the wrapping material) and the actual measured scouring yield is the official conditioned weight, and this serves as the major grounds for pricing wool on a clean-wool basis.

8. *Modernisation of wool technical administration*: This refers to the whole process of flock husbandry, flock shearing, pre-sale clip preparation, i.e. grading and sorting, mechanical baling, mark painting, objective measurement, and wool sample display for sale and pricing wool on a clean-wool basis, etc. This is a general term for the adoption of international scientific administration methods.

9. *Quality count*: This shows the fineness of wool on the basis of the wool fibre diameter (μm). Wool which has a small value of fibre diameter has a higher quality count.

10. *Staple length*: This refers to the staple length (cm) measured from the base to the tip of the staple fibre under natural conditions.

11. *Wool quality grade*: At present wool is classified into three quality grades (semi-fine wool is not considered for the time being) on the basis of the wool types, staple structure, crimpness, growth uniformity, content and colour of grease and suint, and the extent of cleanness after skirting. In addition, a tentative standard sample collection of different wool grades and sorts has been established which serves as the reference for the investigations of the quality classification of domestic wool during the first phase.

12. *Quality differential ratio*: This is also called the price differential ratio and refers to the economic differential ratio on the basis of the actual wool quality (fineness, staple length and quality grades are tentatively regarded as the major items), combined with useful value according to market demand. In this scheme, a wool grade is used as the reference (with a quality differential ratio of 100%), and the quality differential ratio of different grades can then

be calculated. For example, according to the base wool price stipulations, the grade of "64B2II" is regarded as the basic grade which has a quality differential ratio of 100%. Therefore, the quality differential ratio of wool meeting the standard of the grade of a certain quality count (fineness), staple length and quality will be used as the basis in the calculation of the wool price.

13. *Technical error in measurement*: In general, this refers to the measured error resulting from the differences between sampling and instrument operations, or between the technical staff, and the atmosphere humidity and temperature and measure conditions, and other technical factors.

14. *Absolute value*: This refers to the actual value without a comparative object. It is the absolute value of the error of appraisal items expressed in percentage, i.e. it is not related to the percentage of the original appraisal results.

15. *Range of claims*: This refers, in general, to exceeding the range of claim between the result of check-and-accept upon the wool arrival and the error of the original measurement certificate. For example, the conditioned weight of clean wool exceeds 2%; the wool fineness exceeds the lower limit of the particular quality count, staple length exceeds the lower limit of the particular sort and the content of vegetable matter exceeds 2%. It is different from the allowed technical errors in measurement. For example, the content of vegetable matter which had been examined before shipment, was 1.4%, and the result of joint sampling and re-inspection was 1.8% with an error of 0.4%. This has already exceeded the allowed technical error and shows that the previous inspection is questionable. However, the actual content of vegetable matter of 1.8% is still well below 2%. There will be no claim at all because it does not exceed the range of claim and should be considered qualified wool.

16. *Names of defect wool*:

(1)	Seedy and thorny wool:	Wool contains grass seeds, and seeds of *Xanthium*, puncture vine, etc.
(2)	Contaminated piece wool:	Wool with dung and other contaminated materials and difficult to split.
(3)	Skin-piece wool:	Wool with skin pieces owing to careless shearing operation.
(4)	Yellow-stained wool:	Wool which has become yellow in colour owing to dampness or dung and urine contamination.
(5)	Pitch wool:	Wool with traces of pitch markings.
(6)	Paint wool:	Wool with traces of paint markings.
(7)	Scabby wool:	Wool shorn from sheep which have suffered from scabby disease. It also has scab or skin scraps.
(8)	Tender wool:	Some part of the wool diameter becomes thin because of malnutrition or disease.
(9)	Cotted wool:	Wool fibre which has cotted together or formed patches or bundles owing to dampness, rubbing or some other factor.

Appendix H

(Attachment)

Quality Differential Ratio of Different Classification of Fine Wool (I) (%)

Quality grade:				I			
Staple length:		A	B1	B2	C1	C2	D
Fineness	70s	122	117	112	101	84	62
	66s	118	112	108	97	81	59
	64s	113	108	104	94	78	57
	60s	109	104	100	90	75	55

Quality Differential Ratio of Different Classification of Fine Wool (II) (%)

Quality grade:				II			
Staple length:		A	B1	B2	C1	C2	D
Fineness	70s	118	112	108	97	81	59
	66s	113	108	104	94	78	57
	64s	109	104	100	90	75	55
	60s	105	100	96	86	72	53

Quality Differential Ratio of Different Classification of Fine Wool (III) (%)

Quality grade:				III			
Staple length:		A	B1	B2	C1	C2	D
Fineness	70s	114	109	105	94	79	58
	66s	110	105	101	91	76	55
	64s	106	101	97	87	73	53
	60s	102	97	93	84	70	51

Quality Differential Ratio of Improved Wool

Quality grade	Quality differential ratio
I	86%
II	62%
III	48%
Lower than III	33%

Quality Differential Ratio of Edge-Trimmed Fine Wool

Quality count	Length grade higher than B	Length grade lower than C
64s and higher	7.5%	6.5%
60s	7.0%	6.0%

Quality Differential Ratio of Edge-Trimmed Improved Wool and Other Wool

Quality grade (or type)	Quality differential ratio
Edge-trimmed wool higher than Grade II	53%
Edge-trimmed wool lower than Grade III	40%
Head, leg and tail wool	40%
Autumn–summer wool	50%
Coloured and spotted wool, raw-skin clip wool	60%

References

An, X. (1989) The development and improvement of agricultural marketing in China. In: Longworth, J.W. (ed.), *China's Rural Development Miracle*. The University of Queensland Press, St. Lucia, pp.18–25.

Anderson, K. and Park, Y. (1988) China and the international relocation of world textile and clothing activity. *Pacific Economic Paper* 158. Australia–Japan Research Centre, Australian National University, April.

Annual Report Group of the Rural Economy (ARGRE) (1994) *The Blue Book of Annual Economic Development 1993: Agriculture*. China's Social Science Publications, Beijing.

Anon. (1989) Immediate calling for pasture preservation in China. *Economic Daily,* 18 October, p.4.

Brown, C.G. (1993) *Report on Australian Mission to China on Wool Processing in the Pastoral Region*. Department of Agriculture, The University of Queensland, St. Lucia.

Brown, C.G. (1995) Reforming China's agribusiness sector: the case of wool auctions. *China Economic Review* (in preparation).

Brown, C.G. and Longworth, J.W. (1992) Reconciling national economic reforms and local investment decisions in China: fiscal decentralisation and first-stage wool processing. *Development Policy Review* 10, 389–402.

Brown, C.G. and Longworth, J.W. (1994) Lifting the wool curtain: recent reforms and new opportunities in the Chinese wool market with special reference to the "up-country" mills. *Review of Marketing and Agricultural Economics* 62(3), 1–19.

Chen, J. and Lin, X. (1994) Changing role of supply and marketing cooperatives in China's wool marketing. Paper presented to a workshop on The Wool Industry in China at the 38th Annual Conference of the Australian Agricultural Economics Society, Wellington, New Zealand, February.

Chen, L.Y. and Buckwell, A. (1991) *Chinese Grain Economy and Policy*. CAB International, Wallingford.

Cheng, E. (1991) Trouble at the mill. *Far Eastern Economic Review,* 7 March, p.56.

Chinese Members of ACIAR 8811 Project (eds.) (1994) *Economic Aspects of Raw Wool Production and Marketing in China*. Agricultural Scientech Publishing House of China, Beijing (in Chinese).

Connolly, G.P. and Roper, H.E. (1991) China's wool marketing system and the demand for wool. *Agriculture and Resources Quarterly* 3(3), 371–382.

Davis, D. (1988) Unequal chances, unequal outcomes: pension reform and urban inequality. *The China Quarterly* 114, 223–242.

Davis-Friedmann, D. (1983) *Long Lives*. Harvard University Press, Cambridge, MA.

Editorial Board of the Almanac (1990 onwards) *Almanac of China's Textile Industry*. China Textile Press, Beijing.

Editorial Board of China Agriculture Yearbook (1994 and earlier) *China Agriculture Yearbook*. Agricultural Publishing House, Beijing.

Findlay, C. (1992) *Challenges of Economic Reform and Industrial Growth: China's Wool War*, Allen and Unwin, Sydney.

Findlay, C. and Li, Z. (1992) Chinese wool textile industry growth and the demand for raw wool. In: Findlay, C. (ed.), *Challenges of Economic Reform and Industrial Growth: China's Wool War.* Allen and Unwin, Sydney, pp.97–120.

Francis, P. (1992) Fibre diameter distribution: scientific acceptance now emerging. *Farm Journal,* December, 42–57.

Garnaut, R., Bennett, S. and Price, R. (1993) *Wool: Structuring for Global Realities: Report of the Wool Industry Review Committee.* Department of Primary Industry and Energy, Canberra.

Gerritsen, R. (1992) Labour's final rural "crisis"?: Australian rural policy in 1990 and 1991. *Review of Marketing and Agricultural Economics* 60(2), 95–112.

Harmsworth, T. and Day, G. (1990) *Wool and Mohair.* Inkata Press, Melbourne.

Huang, X. (1991) Zhu prescribes a bitter pill for textile recovery. *China Daily,* 15 September, p.1.

International Wool Secretariat (IWS) (1994) *CHINA 2000: Opportunities for Wool Textile Products.* Wool Development International Ltd, Ilkley, West Yorkshire.

Lehane, R. (1993) China and wool—a giant constrained. In: *Partners in Research for Development No. 6* Australian Centre for International Agricultural Research, Canberra, pp.2–9.

Lin, J.Y. (1992) Rural reforms and agricultural growth in China. *American Economic Review* 82(1), 34–51.

Lin, X. (1990) Development of the pastoral areas of Chifeng City Prefecture. In: Longworth, J.W. (ed.), *The Wool Industry in China: Some Chinese Perspectives.* Inkata Press, Melbourne.

Lin, X. (1993) The outlook for Chinese wool production and marketing: some policy proposals. In: Longworth, J.W. (ed.), *Economic Aspects of Raw Wool Production and Marketing,* ACIAR Technical Report No. 25. Australian Centre for International Agricultural Research, Canberra, pp.50–57.

Liu, Y. (1993) Rangeland degradation in the pastoral regions of China: causes and countermeasures. In: Longworth, J.W. (ed.), *Economic Aspects of Raw Wool Production and Marketing,* ACIAR Technical Report No. 25. Australian Centre for International Agricultural Research, Canberra, pp.20–26.

Longworth, J.W. (ed.) (1989) *China's Rural Development Miracle: With International Comparisons.* University of Queensland Press, St. Lucia.

Longworth, J.W. (ed.) (1990) *The Wool Industry in China: Some Chinese Perspectives.* Inkata Press, Melbourne.

Longworth, J.W. (1993) Wool marketing in China: a system in transition. In: Longworth, J.W. (ed.), *Economic Aspects of Raw Wool Production and Marketing,* ACIAR Technical Report No. 25. Australian Centre for International Agricultural Research, Canberra, pp.36–49.

Longworth, J.W. and Williamson, G.J. (1993) *China's Pastoral Region: Sheep and Wool, Minority Nationalities, Rangeland Degradation, and Sustainable Development.* CAB International, Wallingford.

Longworth, J.W. and Williamson, G.J. (1995) *Fine Wool Sheep Breeds and Breeding in China: The Impact of the Australian Merino.* (in preparation).

Martin, W. (1992) Effects of foreign exchange reform on raw wool demand: a quantitative analysis. In: Findlay, C. (ed.), *Challenges of Economic Reform and Industrial Growth: China's Wool War.* Allen and Unwin, Sydney.

Morris, P., Roper, H., Short, C., Proctor, W. and Connolly, G. (1993) *China's Wool Textile Industry: Strategies for Growth.* ABARE Research Report No. 93.9. Australian Bureau of Agricultural and Resource Economics, Canberra.

Parish, W.L. (ed.) (1985) *Chinese Rural Development.* M.E. Sharp Inc., New York.

Power, J. (1993) China reneges on wool deals in price crisis. *Australian Financial Review,* 23 April, p.3.

Shaffer, J.D. and Wen, S. (1994) The transformation from low income agricultural economies: observations focusing on Africa and China. Invited paper presented to the XXII International Conference of Agricultural Economists, Harare, Zimbabwe, August.

Shi, Z. (1990) Wool marketing problems in China. In: J.W. Longworth (ed.), *The Wool Industry in China: Some Chinese Perspectives*. Inkata Press, Melbourne.

State Statistical Bureau of the PRC (1991 onwards) *China Statistical Yearbook*. China Statistical Information and Consultancy Service Center, Beijing.

Sun, W. (1992) Beijing encourages laid-off workers to explore service jobs. *China Daily* 6 June, p.1.

Watson, A. (1983) Agriculture looks for "shoes that fit": the production responsibility system and its implications. *World Development* 11(8), 705–730.

Watson, A. (1988) The reform of agricultural marketing in China since 1978. *The China Quarterly* 113, 1–28.

Watson, A. (1989) Investment issues in the Chinese countryside. *The Australian Journal of Chinese Affairs* 22, 85–126.

Watson, A. (1994) *Market Reform and Agricultural Development in China* Working Paper No. 94/3. Chinese Economy Research Unit, University of Adelaide, Adelaide.

Watson, A. and Findlay, C. (1992) The "wool war" in China. In: Findlay, C. (ed.), *Challenges of Economic Reform and Industrial Growth: China's Wool War*. Allen and Unwin, Sydney, pp.163–180.

Watson, A., Findlay, C. and Du Yintang (1989) Who won the wool war?: a case study of rural product marketing in China. *The China Quarterly* 118, 213–41.

Wen-hui, T. (1987) Life after retirement: elderly welfare in China. *Asian Survey* XXVII(5), 566–576.

Wilcox, C. (1994) Wool importing system of China. Paper presented to a workshop on The Wool Industry in China at the 38th Annual Conference of the Australian Agricultural Economics Society, Wellington, New Zealand, February.

World Bank (1988) *China: Finance and Investment*. World Bank, Washington, DC.

Wu, Y. (1993) *One Industry, Two Regimes: The Chinese Textile Sector Growth, Reforms and Efficiency*, Working Paper No. 93/2. Chinese Economy Research Unit, University of Adelaide, Adelaide.

Wu, Z. (1990) Chinese wool market—then and now. *Wool Technology and Sheep Breeding* XXXVIII(2), 41–43.

Zhai, F. (1991) Textile firms moving to end slump. *China Daily*, 13 October, p.4.

Zhang, C. (1990a) An overview of the Chinese wool textile industry. In: Longworth, J.W. (ed.), *The Wool Industry in China: Some Chinese Perspectives*. Inkata Press, Melbourne, pp.31–42.

Zhang, C. (1990b) Review of wool production and wool requirements in China. In: Longworth, J.W. (ed.), *The Wool Industry in China: Some Chinese Perspectives*. Inkata Press, Melbourne, pp.8–23.

Zhang, C. (1993) The development of Chinese wool auctions. In: Longworth, J.W. (ed.), *Economic Aspects of Raw Wool Production and Marketing*, ACIAR Technical Report No. 25. Australian Centre for International Agricultural Research, Canberra, pp.58–62.

Zhang, C. (1994) Review of wool production and wool requirements in China. In: Chinese Team Members of ACIAR 8811 (eds), *Economic Aspects of Raw Wool Production and Marketing in China*. Agricultural Scientech Publishing House of China, Beijing (in Chinese), pp.4–33.

Zhang, C. and Niu, R. (1994) Development of Chinese wool auctions. Paper presented to a workshop on The Wool Industry in China at the 38th Annual Conference of the Australian Agricultural Economics Society, Wellington, New Zealand, February.

Zhou, L. (1990) Economic development in China's pastoral regions. In: J.W. Longworth (ed.), *The Wool Industry in China: Some Chinese Perspectives*. Inkata Press, Melbourne, pp.43–56.

Index